McGraw-Hill
My Math

Welcome to My Math — your very own math book!
You can write in it — in fact, you are encouraged to write, draw, circle, explain, and color as you explore the exciting world of mathematics. Let's get started. Grab a pencil and finish each sentence.

My name is _____.

My favorite color is _____.

My favorite hobby or sport is _____.

My favorite TV program or video game is

_____.

My favorite class is _____.

Math, of course!

McGraw Hill Education

Bothell, WA • Chicago, IL • Columbus, OH • New York, NY

connectED.mcgraw-hill.com

 **Education**

Copyright © 2013 The McGraw-Hill Companies, Inc.

STEM McGraw-Hill is committed to providing instructional materials in Science, Technology, Engineering, and Mathematics (STEM) that give all students a solid foundation, one that prepares them for college and careers in the 21st century.

Send all inquiries to:
McGraw-Hill Education
STEM Learning Solutions Center
8787 Orion Place
Columbus, OH 43240

ISBN: 978-0-02-116191-1 *(Volume 2)*
MHID: 0-02-116191-7

Printed in the United States of America.

14 DOW 19 18 17 16 15

STEM

Our mission is to provide educational resources that enable students to become the problem solvers of the 21st century and inspire them to explore careers within Science, Technology, Engineering, and Mathematics (STEM) related fields.

 The *McGraw-Hill* Companies

Meet The Artists!

Abby Crutchley

The Market Math Winning this contest made me feel like I had butterflies in my stomach! I didn't think I would win out of the 72 finalists. *Volume 1*

Matt Gardner

Using Math to Build Math means everything to me. It is my favorite subject in school. I got the idea because I like to build and I thought a K'Nex vehicle would make a cool math book cover. *Volume 2*

Other Finalists

Jacob Alvarez
Beach Math-1

Emily Jiang
Math Angel

India Johnson
Math in Nature

Nathan Baal
Math in My Neighborhood

Sergio Reyes
Math With My Fingers

Kelsey Thompson
Doggy Shapes

Maddie Mathews
The Book of Math-ews

Kaya Ross
Math is in My Neighborhood

Yonaton Barkel
We Use Math
Skills Everyday

Ellie Hull
Patterns
Qwirkle™ tiles reproduced with the permission of MindWare®.

Find out more about the winners and other finalists at www.MHEonline.com.

We wish to congratulate all of the entries in the 2011 *McGraw-Hill My Math* "What Math Means To Me" cover art contest. With over 2,400 entries and more than 20,000 community votes cast, the names mentioned above represent the two winners and ten finalists for this grade.

GO digital

it's all at
connectED.mcgraw-hill.com

Go to the Student Center for your eBook, Resources, Homework, and Messages.

Write your Username [   ] Password [   ]

Get your resources online to help you in class and at home.

Vocab

Find activities for building vocabulary.

Watch

Watch animations of key concepts.

Tools

Explore concepts with virtual manipulatives.

Check

Self-assess your progress.

eHelp

Get targeted homework help.

Games

Reinforce with games and apps.

Tutor

See a teacher illustrate examples and problems.

GO mobile

Scan this QR code with your smart phone* or visit mheonline.com/stem_apps.

*May require quick response code reader app.

Available on the App Store

v

Contents in Brief

Organized by Domain

Common Core State Standards

Standards for Mathematical PRACTICE  Woven Throughout

Chapter

1

Place Value

ESSENTIAL QUESTION
How can numbers be expressed, ordered, and compared?

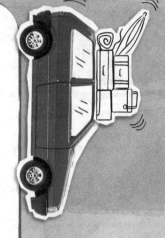

Look for this! Watch ▶

Click online and you can watch videos that will help you learn the lessons.

connectED.mcgraw-hill.com

Chapter

2 Addition

ESSENTIAL QUESTION
How can place value help me add larger numbers?

connectED.mcgraw-hill.com

Chapter

3 Subtraction

ESSENTIAL QUESTION
How are the operations of subtraction and addition related?

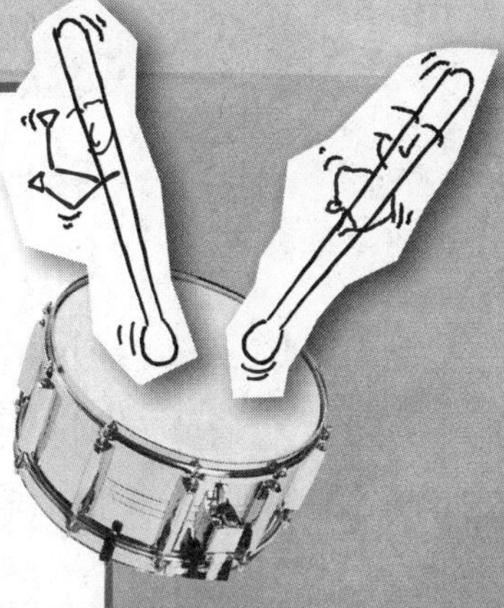

Getting Started

Lessons and Homework

Wrap Up

Look for this!
Click online and you can get more help while doing your homework.

eHelp

connectED.mcgraw-hill.com

Chapter 4
Understand Multiplication

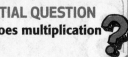

ESSENTIAL QUESTION
What does multiplication mean?

Getting Started

Lessons and Homework

Wrap Up

Yummy!

connectED.mcgraw-hill.com

x

Chapter 5 Understand Division

Getting Started

Lessons and Homework

Wrap Up

connectED.mcgraw-hill.com

Look for this!
Click online and you can find tools that will help you explore concepts.

Chapter 6 Multiplication and Division Patterns

ESSENTIAL QUESTION
What is the importance of patterns in learning multiplication and division?

BLAST OFF!

connectED.mcgraw-hill.com

Chapter 7 Multiplication and Division

ESSENTIAL QUESTION
What strategies can be used to learn multiplication and division facts?

Getting Started

Lessons and Homework

Wrap Up

Look for this!

Click online and you can watch a teacher solving problems.

Tutor

connectED.mcgraw-hill.com

Chapter **8** Apply Multiplication and Division

ESSENTIAL QUESTION
How can multiplication and division facts with smaller numbers be applied to larger numbers?

Getting Started

Lessons and Homework

Wrap Up

connectED.mcgraw-hill.com

Chapter 9

Properties and Equations

ESSENTIAL QUESTION
How are properties and equations used to group numbers?

Getting Started

Lessons and Homework

Wrap Up

connectED.mcgraw-hill.com

Look for this!
Click online and you can find activities to help build your vocabulary.

Vocab

Chapter 10 Fractions

ESSENTIAL QUESTION
How can fractions be used to represent numbers and their parts?

Getting Started

Lessons and Homework

Wrap Up

I've got the answer!

connectED.mcgraw-hill.com

Chapter 11 Measurement

Getting Started

Lessons and Homework

Wrap Up

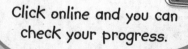

Look for this! Check ✓

Click online and you can check your progress.

connectED.mcgraw-hill.com

Chapter 12 Represent and Interpret Data

ESSENTIAL QUESTION
How do we obtain useful information from a set of data?

connectED.mcgraw-hill.com

Chapter 13 Perimeter and Area

ESSENTIAL QUESTION
How are perimeter and area related and how are they different?

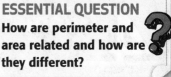

I dig this area!

Getting Started

Lessons and Homework

Wrap Up

connectED.mcgraw-hill.com

Chapter 14 Geometry

ESSENTIAL QUESTION
How can geometric shapes help me solve real-world problems?

Getting Started

Lessons and Homework

Wrap Up

connectED.mcgraw-hill.com

Chapter

10 Fractions

A Day at My School

ESSENTIAL QUESTION

How can fractions be used to represent numbers and their parts?

Watch a video!

Watch

MY Common Core State Standards

Number and Operations – Fractions

3.NF.1 Understand a fraction $\frac{1}{b}$ as the quantity formed by 1 part when a whole is partitioned into b equal parts; understand a fraction $\frac{a}{b}$ as the quantity formed by a parts of size $\frac{1}{b}$.

3.NF.2 Understand a fraction as a number on the number line; represent fractions on a number line diagram.

3.NF.2a Represent a fraction $\frac{1}{b}$ on a number line diagram by defining the interval from 0 to 1 as the whole and partitioning it into b equal parts. Recognize that each part has size $\frac{1}{b}$ and that the endpoint of the part based at 0 locates the number $\frac{1}{b}$ on the number line.

3.NF.2b Represent a fraction $\frac{a}{b}$ on a number line diagram by marking off a lengths $\frac{1}{b}$ from 0. Recognize that the resulting interval has size $\frac{a}{b}$ and that its endpoint locates the number $\frac{a}{b}$ on the number line.

3.NF.3 Explain equivalence of fractions in special cases, and compare fractions by reasoning about their size.

3.NF.3a Understand two fractions as equivalent (equal) if they are the same size, or the same point on a number line.

3.NF.3b Recognize and generate simple equivalent fractions, e.g., $\frac{1}{2} = \frac{2}{4}, \frac{4}{6} = \frac{2}{3}$. Explain why the fractions are equivalent, e.g., by using a visual fraction model.

3.NF.3c Express whole numbers as fractions, and recognize fractions that are equivalent to whole numbers.

3.NF.3d Compare two fractions with the same numerator or the same denominator by reasoning about their size. Recognize that comparisons are valid only when the two fractions refer to the same whole. Record the results of comparisons with the symbols >, =, or <, and justify the conclusions, e.g., by using a visual fraction model.

Geometry *This chapter also addresses this standard:*

3.G.2 Partition shapes into parts with equal areas. Express the area of each part as a unit fraction of the whole.

Standards for Mathematical PRACTICE

1. Make sense of problems and persevere in solving them.
2. Reason abstractly and quantitatively.
3. Construct viable arguments and critique the reasoning of others.
4. Model with mathematics.
5. Use appropriate tools strategically.
6. Attend to precision.
7. Look for and make use of structure.
8. Look for and express regularity in repeated reasoning.

= focused on in this chapter

Name _____

Am I Ready?

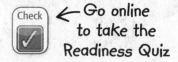

Go online to take the Readiness Quiz

Write the number of parts. Tell whether each figure shows parts that are *equal* or *not equal*.

1.

_____

2.

_____

3.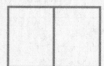

_____

Write the name of the equal parts. Write *halves*, *thirds*, or *fourths*.

4.

_____

5.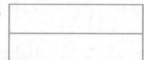

_____

6.

_____

Circle the point which represents each given number.

7. 380

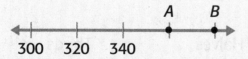

8. 169

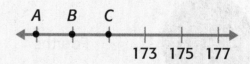

9. Jill draws a figure and divides it into thirds.
Draw what her figure could look like.

Shade the boxes to show the problems you answered correctly.

How Did I Do? | 1 | 2 | 3 | 4 | 5 | 6 | 7 | 8 | 9 |

MY Math Words

Vocab

Review Vocabulary

fourths　　　halves　　　thirds

Making Connections

Circle the review vocabulary word which represents each example.

	Fourths	Halves	Thirds
$\frac{1}{3}$ $\frac{1}{3}$ $\frac{1}{3}$	Fourths	Halves	Thirds
$\frac{1}{4}$ $\frac{1}{4}$ $\frac{1}{4}$ $\frac{1}{4}$	Fourths	Halves	Thirds
$\frac{1}{2}$ $\frac{1}{2}$	Fourths	Halves	Thirds

What does *whole* mean? What does *part of a whole* mean?

_____

_____

MY Vocabulary Cards

Mathematical
PRACTICE

Lesson 10–2

denominator

$$\frac{5}{8}$$

Lesson 10–6

equivalent fractions

$$\frac{2}{4} = \frac{1}{2}$$

Lesson 10–1

fraction

$$\frac{3}{4}$$

Lesson 10–2

numerator

$$\frac{5}{6}$$

Lesson 10–1

unit fraction

$$\frac{1}{3}$$

Ideas for Use

- Write a tally mark on each card every time you read the word in this chapter or use it in your writing. Challenge yourself to use at least 5 tally marks for each card.

- Use the blank cards to draw or write phrases or examples that will help you with concepts like comparing fractions.

Fractions that represent the same part of the whole.

The prefix *non-* means "not." Describe 2 non-equivalent fractions. Write an example.

_____

_____

The bottom number in a fraction. It is the total number of equal parts.

Draw models to show three different fractions that all have a denominator of 4.

The top number in a fraction. It is the number of equal parts being represented.

In the fraction $\frac{4}{8}$, what is the numerator? What does it show?

_____

_____

A number that represents an equal part of a whole or an equal part of a set.

Draw a rectangle, and divide it into 6 equal parts. Color the parts to represent $\frac{3}{6}$.

Represents the quantity formed by one part when a whole is partitioned into equal parts.

Unit can mean "a single thing." How does this help you understand the definition of *unit fraction*?

_____

_____

MY Foldable

FOLDABLES® Follow the steps on the back to make your Foldable.

✂

$\overline{2}$ $\overline{2}$

$\overline{6}$ $\overline{6}$ $\overline{6}$ $\overline{6}$ $\overline{6}$ $\overline{6}$

$\overline{8}$ $\overline{8}$ $\overline{8}$ $\overline{8}$ $\overline{8}$ $\overline{8}$ $\overline{8}$ $\overline{8}$

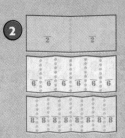

1

$$\overline{3} \qquad \overline{3} \qquad \overline{3}$$

$$\overline{4} \qquad \overline{4} \qquad \overline{4} \qquad \overline{4}$$

Name ....

Number and Operations–Fractions
3.NF.1, 3.G.2

CCSS

Unit Fractions

Lesson 1

ESSENTIAL QUESTION
How can fractions be used to represent numbers and their parts?

A **fraction** is a number that represents an equal part of a whole or equal part of a set.

 Math in My World Tools Watch Tutor

One for you! One for me!

Example 1

Danny shared his granola bar with Steph. He broke it into two equal pieces. What fraction of the granola bar did each receive?

Model the whole granola bar.

Place two equal-sized fraction tiles so that the combined length is equal to one whole.

1

What fraction tiles did you place? ⬜/⬜ tiles

So each person received ⬜/⬜, or one half, of a whole granola bar.

A **unit fraction** represents one equal part of the whole. The top number of a unit fraction is 1.

Example 2

One whole is divided into four equal parts. What unit fraction represents one equal part of the whole?

Write the unit fraction.

1 ← one part

⬜ ← The whole was partitioned into four equal parts.

1

$\frac{1}{4}$ $\frac{1}{4}$ $\frac{1}{4}$ $\frac{1}{4}$

The unit fraction is ⬜/⬜, or one-fourth.

Example 3

Ian made a loaf of bread as his final project in health class. He divided the loaf equally among some students. Each student received $\frac{1}{8}$ of the loaf. Into how many equal parts did Ian cut the loaf?

Use the fraction tiles [**1**] and [$\frac{1}{8}$]

to model one whole and one whole divided into equal parts. Draw your models.

> *My Drawing!*

Count the equal parts. There are _____ equal parts.

Label each equal-sized part $\frac{1}{8}$. ◄ [one part of eight or *one-eighth*]

So, Ian cut the loaf into _____ equal parts, or eighths.

Guided Practice

Divide the whole into equal parts. Then label each part with its unit fraction.

1. 2 equal parts

[ ]

2. 4 equal parts

[ ]

3. 8 equal parts

[ ]

Talk MATH

What is a unit fraction?

Independent Practice

Divide the whole into equal parts. Then label each part with its unit fraction.

4. 3 equal parts

5. 6 equal parts

Write how many equal parts. Shade one part. Write its unit fraction.

6.

_____ equal parts

unit fraction: ⬜／⬜

7.

_____ equal parts

unit fraction: ⬜／⬜

8.

_____ equal parts

unit fraction: ⬜／⬜

9.

_____ equal parts

unit fraction: ⬜／⬜

Circle the correct unit fraction for the shaded section of each model.

10.

$\frac{1}{3}$ $\frac{1}{4}$

11.

$\frac{1}{8}$ $\frac{1}{6}$

12.

$\frac{1}{5}$ $\frac{1}{6}$

13.

$\frac{1}{2}$ $\frac{1}{3}$

14.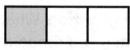

$\frac{1}{3}$ $\frac{1}{4}$

15.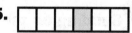

$\frac{1}{5}$ $\frac{1}{6}$

Problem Solving

16. Max folded a piece of paper in half. Then he folded it in half one more time. How many equal-sized parts did he have when he opened the paper? What unit fraction represents each part?

17. Mathematical **PRACTICE** 5 **Use Math Tools** Jenny is holding a fraction tile labeled $\frac{1}{3}$. How many $\frac{1}{3}$-fraction tiles are needed to equal the fraction tile labeled 1?

18. Mr. Clinger divided the gymnasium floor into 8 equal sections. Draw lines to divide the whole into equal parts. Then label each part with its unit fraction.

19. Mathematical **PRACTICE** 2 **Reason** How are all unit fractions alike? How are they different?

20. **Building on the Essential Question** What happens to the size of each equal part when you divide a whole into more and more equal parts?

Name _____

MY Homework

Homework Helper

Need help? connectED.mcgraw-hill.com

Franny wants to braid string to make a bracelet. She has one long piece of string. Franny needs to divide the string into 3 equal pieces for braiding. Model the whole string divided into 3 pieces. Write the unit fraction for 1 piece of the string.

1. Use the one whole fraction tile to represent the whole piece of string.

2. Use $\frac{1}{3}$-fraction tiles to model 3 equal parts.

3. The string was divided into 3 equal pieces. The unit fraction that represents 1 of those pieces is $\frac{1}{3}$.

Practice

Divide the whole into equal parts. Then label each part with its unit fraction.

1. four equal parts

2. two equal parts

3. six equal parts

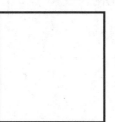

4. three equal parts

Write how many equal parts. Shade one part. Write its unit fraction.

5. _____ equal parts

unit fraction: $\dfrac{\Box}{\Box}$

6. _____ equal parts

unit fraction: $\dfrac{\Box}{\Box}$

7.  _____ equal parts

unit fraction: $\dfrac{\Box}{\Box}$

8.  _____ equal parts

unit fraction: $\dfrac{\Box}{\Box}$

 ## Problem Solving

9. **Mathematical PRACTICE ③ Justify Conclusions** Louis has a rectangular piece of construction paper. Can he divide the shape into 4 equal parts? Explain.

_____

_____

Vocabulary Check

Choose the correct word(s) to complete each sentence.

fraction   unit fraction

10. A _____ is exactly one equal part of a whole.

11. A _____ represents an equal part of a whole.

Test Practice

12. Which unit fraction represents the shaded part of the whole?

Ⓐ $\dfrac{1}{3}$ Ⓒ $\dfrac{1}{6}$

Ⓑ $\dfrac{1}{4}$ Ⓓ $\dfrac{1}{8}$

Part of a Whole

Lesson 2

ESSENTIAL QUESTION
How can fractions be used to represent numbers and their parts?

Math in My World

Tools   Watch   Tutor

Example 1

The background paper of Mrs. Dempsey's bulletin board is divided equally into four stripes. What fraction of the bulletin board is green?

Use a model.

The bulletin board represents one whole.
The whole is divided into 4 equal parts, or fourths.
Use fraction tiles to model the equal parts of the bulletin board. Trace the tiles. Color one part green.

Write: $\frac{1}{4}$

Read: *one-fourth*

1

| $\frac{1}{4}$ | $\frac{1}{4}$ | $\frac{1}{4}$ | $\frac{1}{4}$ |

Helpful Hint
Sometimes $\frac{1}{4}$ is also read *one-quarter*.

part that is green ⟶ **$\frac{1}{4}$** ⟵ numerator
total number of parts ⟶ ⟵ denominator

The **numerator** tells the number of equal parts being represented.

The **denominator** tells the total number of equal parts.

So, $\frac{1}{4}$, or *one-_____*, of the bulletin board is _____ .

Not all fractions are unit fractions.

Example 2

What fraction of the flag is red?

$\dfrac{2}{3}$ ← red parts
← total number of equal parts

Write: $\dfrac{2}{\boxed{}}$

Read: _____-*thirds*

So, $\dfrac{2}{3}$, or *two-*_____ of the flag is _____.

Explain why $\dfrac{2}{3}$ is not a unit fraction.

_____

_____

Guided Practice

Complete the chart. Write a fraction for each part.

Fraction Model	Part that is Yellow	Part that is *not* Yellow
1.	$\dfrac{}{}$	$\dfrac{}{}$
2.	$\dfrac{}{}$	$\dfrac{}{}$

Talk MATH

What is the difference between the numerator and the denominator of a fraction?

Name _____

Independent Practice

Complete the chart. Write a fraction for each part.

Fraction Model	Part that is Blue	Part that is *not Blue*
3.	☐/☐	☐/☐
4.	☐/☐	☐/☐
5.	☐/☐	☐/☐

6. What fraction of the honeycomb has bees?

☐/☐

7. What fraction of the figure is *not* shaded?

☐/☐

Shade each figure to represent the fraction.

8. $\frac{2}{4}$

9. $\frac{2}{8}$

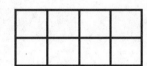

10. $\frac{2}{2}$

Match each fraction to its word name.

11. $\frac{3}{8}$ • *five-sixths*

12. $\frac{5}{6}$ • *three-fourths*

13. $\frac{3}{4}$ • *three-eighths*

Problem Solving

Mathematical PRACTICE **5** **Use Math Tools** The primary colors are red, blue, and yellow. The secondary colors are green, orange, and violet. Use the color wheel to answer Exercises 14–17.

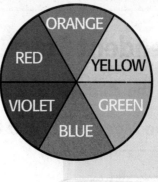

14. What fraction of the primary and secondary colors is red?

☐
☐

15. What fraction is blue or orange?

☐
☐

16. What fraction is not violet?

☐
☐

17. What fraction of colors is not a primary color?

☐
☐

HOT Problems

Mathematical PRACTICE **4** **Model Math** Draw and shade a model to represent the fraction $\frac{3}{4}$.

18.

19. **?** **Building on the Essential Question** Explain how to write a fraction to describe part of a whole.

_____

_____

My Work!

MY Homework

Homework Helper

Need help? connectED.mcgraw-hill.com

Dennis and 2 friends are sharing a submarine sandwich equally. All but one part has hot peppers. What fraction of the sandwich has hot peppers? What fraction of the sandwich does not have hot peppers?

Model the problem. The entire sandwich is the whole. It is divided into 3 equal parts. Two of the 3 parts have hot peppers.

parts with hot peppers ⟶
total number of equal parts ⟶ $\dfrac{2}{3}$

$\dfrac{1}{3}$ ⟵ part without hot peppers
⟵ total number of equal parts

So, $\dfrac{2}{3}$ of the sandwich has hot peppers, and $\dfrac{1}{3}$ of the sandwich does not.

Practice

Complete the chart. Write a fraction for each part.

Fraction Model	Part that is Green	Part that is *not Green*
1.	▢/▢	▢/▢
2.	▢/▢	▢/▢

Shade each figure to represent the fraction.

3. $\frac{4}{6}$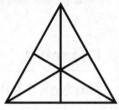

4. $\frac{3}{3}$

5. $\frac{5}{8}$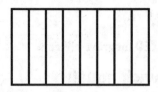

6. $\frac{1}{4}$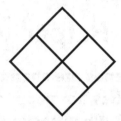

7. **Mathematical PRACTICE** 2 **Stop and Reflect** In Exercises 3–6, circle the unit fraction. Write the fraction below. Explain why it is a unit fraction.

Problem Solving

8. A loaf of bread is cut into 8 equal slices. What fraction of the bread is left after 6 slices have been used for sandwiches?

9. Kristen made a pinwheel with 6 points. She colored 1 point red, 2 points blue, and 3 points purple. What fraction of the points are neither red nor purple?

Vocabulary Check

Draw a line to match the vocabulary term with its meaning.

10. denominator   • the number of parts being represented

11. numerator   • the total number of equal parts

Test Practice

12. Which fraction of the figure is yellow?

Ⓐ $\frac{2}{8}$ Ⓑ $\frac{2}{6}$ Ⓒ $\frac{1}{2}$ Ⓓ $\frac{3}{6}$

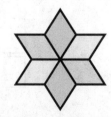

Name ....

Part of a Set

Lesson 3

ESSENTIAL QUESTION
How can fractions be used to represent numbers and their parts?

Fractions can also be used to name part of a set.

 Math in My World Tools Watch Tutor

Example 1

Mrs. Maynard gave each group of students 2 yellow and 4 red markers. What fraction of the set of markers is yellow?

1. Use a set of 6 counters to represent the set of 6 markers.

2 yellow markers   4 red markers

2. Draw and color the counters on the fraction below to represent the fraction of yellow counters.

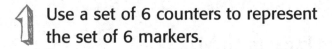  numerator-2 yellow counters

 denominator-6 total counters

$\dfrac{2}{6}$

Write: $\dfrac{2}{6}$ ← use numbers
Read: *two-sixths* ← use words

So, the fraction of yellow markers is .

Online Content at ⌐ connectED.mcgraw-hill.com

Example 2

Cody told his four dogs to sit. What fraction of the set of dogs listened to Cody?

The yellow counters represent the sitting dogs. The red counter represents the standing dog.

3 dogs, out of a total of _____ dogs, are sitting.

Write: $\dfrac{3}{4}$ ← [dogs sitting]
    ← [total dogs]

Read: *three-fourths*

So, $\dfrac{\square}{\square}$ or _____ *-fourths,* of the dogs obeyed.

What fraction of the set of dogs did *not* listen to Cody?

_____ dog, out of a total of 4 dogs, is standing.

So, $\dfrac{\square}{\square}$ or *one-* _____ , of the dogs did *not* listen to Cody.

Guided Practice [Check ✓]

Talk MATH

How is finding the fraction of a set different than finding the fraction of one whole?

Complete the chart. Write a fraction for each part.

Fraction Model	Part that is Yellow	Part that is *not* Yellow
1. 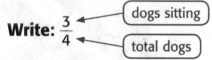	$\dfrac{\square}{\square}$	$\dfrac{\square}{\square}$
2.	$\dfrac{\square}{\square}$	$\dfrac{\square}{\square}$

Independent Practice

Write each fraction.

3. What fraction of the set of daisies is yellow?

 ▢/▢

4. What fraction of the set of buttons is round?

 ▢/▢

5. What fraction of the set of chairs is *not* blue?

 ▢/▢

6. What fraction of the set of seashells is purple?

 ▢/▢

Shade each set to represent the fraction.

7. $\frac{1}{2}$ shaded

8. $\frac{3}{4}$ shaded

9. $\frac{4}{6}$ shaded

10. $\frac{5}{8}$ shaded

Write the missing numerator or denominator.

11. What fraction of the set of spools of thread is red?

 ▢/3

12. What fraction of the set of cups is yellow?

 3/▢

Problem Solving

Use the picture of the fruit to answer Exercises 13–16.

13. There are 8 pieces of fruit in the set. Circle the word used to describe all 8 pieces of fruit.

    numerator   denominator

14. What fraction of the set of fruit is *not* an apple?

15. Suppose Kyle ate one pear. What fraction of the pears did Kyle eat?

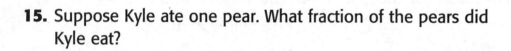

16. PRACTICE 2 **Reason** Describe which pieces of fruit can be represented by the fraction $\frac{5}{8}$.

_____

HOT Problems

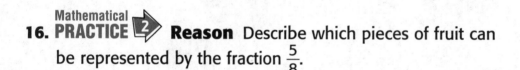

17. PRACTICE 4 **Model Math** Draw a set of objects that represents a fraction with a numerator of 4. Write the fraction.

18. **Building on the Essential Question** How is finding a fractional part of a set different than finding a fractional part of a whole?

_____

_____

MY Homework

Homework Helper

Need help? connectED.mcgraw-hill.com

Carolyn has put together gift bags for guests at her party. There are 6 bags in all. What fraction of the set of bags is yellow? What fraction of the set of bags is blue?

The total number of bags is 6. This is the denominator. The numerator for each fractional part is the number of yellow bags and the number of blue bags.

$\dfrac{4}{6}$ ← yellow bags
← total number of bags

$\dfrac{2}{6}$ ← blue bags
← total number of bags

So, $\dfrac{4}{6}$ of the gift bags are yellow, and $\dfrac{2}{6}$ of the gift bags are blue.

Practice

Shade each set to represent the fraction.

1. $\dfrac{3}{4}$ ◯◯◯◯

2. $\dfrac{4}{6}$ ◯◯◯◯◯◯

3. $\dfrac{2}{3}$ ◯◯◯

4. $\dfrac{1}{2}$ ◯◯

5. Write a fraction for each part.

 $\dfrac{}{}$ part that is red $\dfrac{}{}$ part that is *not* red

6. What fraction of the set of balloons is green?

7. What fraction of the set of books is blue?

8. What fraction of the set of bees is flying away?

9. What fraction of the set of signs is square?

 ## Problem Solving

10. Ramona writes each letter of her first name on separate index cards. What fraction of the cards has a vowel?

11. **Mathematical PRACTICE 1** **Keep Trying** Bryan has 3 nickels, 3 dimes, and 2 quarters. What fraction of the coins is either a dime or a quarter?

12. The Morse family went shoe shopping. Harry got a pair of rain boots and a pair of tennis shoes. Kate got a pair of tennis shoes and a pair of sandals. What fraction of the set of new shoes is rain boots?

Test Practice

13. What fraction of the birds are on the window sill?

Ⓐ $\frac{1}{2}$ Ⓒ $\frac{4}{8}$

Ⓑ $\frac{3}{8}$ Ⓓ $\frac{5}{8}$

Number and Operations – Fractions
3.NF.1

CCSS

Problem-Solving Investigation

STRATEGY: Draw a Diagram

Lesson 4

ESSENTIAL QUESTION
How can fractions be used to represent numbers and their parts?

Learn the Strategy

Anessa and a classmate have 8 insects in a jar. Four-eighths of the insects are beetles. One is a firefly and the rest are crickets. What fraction of the insects are crickets?

1 Understand

What facts do you know?

There are 8 insects. One is a _____.

Four-eighths are _____. The rest are _____.

What do you need to find?

the fraction of insects that are _____

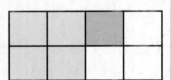

2 Plan

I will draw a diagram to solve the problem.

3 Solve

- Draw a figure that is divided into 8 equal parts.
- Shade $\frac{4}{8}$ of the figure for the beetles. Label with B.
- Shade 1 part for the firefly. Label with F.

So, $\dfrac{}{}$ of the insects are crickets.

4 Check

Does your answer make sense? Explain.

Online Content at connectED.mcgraw-hill.com

Lesson 4   587

Practice the Strategy

Six students brought their pet to school on Pet Day. Three of the pets were dogs and $\frac{1}{6}$ were cats. The rest of the pets were birds. What fraction of the pets were birds?

 Understand

What facts do you know?

_____

_____

What do you need to find?

_____

_____

Plan

_____

_____

Solve

Check

Does your answer make sense? Explain.

_____

_____

Name

Apply the Strategy

Solve each problem by drawing a diagram.

1. Ali is playing jacks. She tosses 8 jacks on the floor. She bounces the ball and picks up 5 jacks before the ball drops. What fraction of the jacks did Ali *not* pick up?

2. Of the 4 houses on the block, 2 are brick and 1 is wood. What fraction of houses is neither wood nor brick?

3. Two out of three students in the Reading Club wear glasses. Write the fraction in words to describe the set of students who wear glasses.

4. There are 6 books. Three-sixths of the books are Willow's. One belongs to Brian. The other books belong to Mrs. Peterson. How many books belong to Mrs. Peterson?

5. **Mathematical PRACTICE 6** **Be Precise** There are an equal number of student desks in classrooms 3A and 3B. Joel washed one-half of the desks in room 3A. Tyrell washed three-fourths of the desks in room 3B. Which boy washed more desks? Explain.

Review the Strategies

Use any strategy to solve each problem.

- Draw a diagram.
- Make a table.
- Look for a pattern.
- Use models.

6. Write the fraction for the shaded part of each figure. How are the fractions for these divided circles alike?

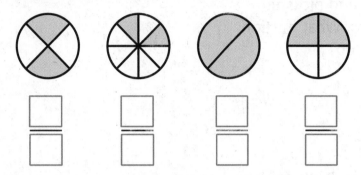

_____

7. Matthew downloaded six songs from the Internet. One-half of the songs were country songs. How many songs were _not_ country songs?

_____

8. Beth gets on the elevator at the sixth floor. She rides up three floors to meet Doris. They ride down seven floors to meet Julio. How many floors is Beth from where she started?

_____

9. **Mathematical PRACTICE 1** **Make Sense of Problems** Four students are standing in line. Evan is ahead of Kam. Chad is behind Kam. Tara is behind Evan. Write the order in which they are standing from front to back in line.

_____

10. There are 8 paintbrushes. One-half of the paintbrushes are red and the rest of the paintbrushes are green. How many paintbrushes are green?

_____

My Work!

MY Homework

Homework Helper

Need help? connectED.mcgraw-hill.com

Alisha is making a bracelet with 4 beads and 4 silver charms. Two of the beads are blue. The rest of the beads are green. What fraction of the bracelet is made up of green beads?

1 Understand

What facts do you know?

Alisha has 4 beads and 4 charms.

Two of the beads are blue.

What do you need to find?

the fraction of the bracelet that is made up of green beads

2 Plan

Draw a diagram to solve the problem.

3 Solve

First, draw a figure divided into 8 equal parts.

S	S	B	B
S	S		

Mark 4 of the parts S for silver charms.

Mark 2 of the parts B for blue beads.

There are 2 parts not filled. Two parts out of 8 parts is $\frac{2}{8}$.

So, the green beads are $\frac{2}{8}$ of the bracelet.

4 Check

Does the answer make sense?

Yes. 4 silver charms + 2 blue beads + 2 green beads = 8 parts.

Problem Solving

Solve each problem by drawing a diagram.

1. Dalton ate 6 pieces of fruit on Monday. He ate 2 apples, 1 banana, 1 orange, and some apricots. What fraction of the fruit Dalton ate was apricots?

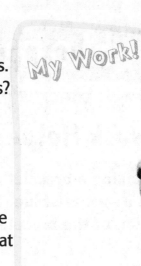

My Work!

2. The Johnsons have 3 dogs. Two of the dogs have brown spots. The other dog has black spots. What fraction of the dogs has black spots?

3. **Mathematical**
   PRACTICE 4 **Model Math** Hannah has 6 cups. She wants to divide them evenly between 2 shelves. How many cups will Hannah put on each shelf?

4. Kenley collects stuffed animals. She has eight stuffed animals. One animal is a bear. What fraction of her stuffed animals is *not* a bear?

5. **Mathematical**
   PRACTICE 2 **Reason** Finn walks 1 mile to the grocery store. When he is halfway back home from the store, a friend picks him up and drives him the rest of the way. What fraction of Finn's round-trip to the grocery store did he *not* walk?

Check My Progress

Vocabulary Check

Choose the correct word(s) to complete each sentence.

denominator   fraction

numerator   unit fraction

1. A _____ is a number that represents an equal part of a whole or an equal part of a set.

2. Exactly one equal part of a whole is called a _____.

3. The _____ tells the number of equal parts being represented.

4. The _____ tells the total number of equal parts.

Concept Check

Divide the whole into equal parts. Then label each part with its unit fraction.

5. 6 equal parts

6. 3 equal parts

Circle the correct unit fraction for the shaded section of each model.

7.

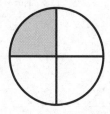

$\frac{1}{3}$ $\frac{1}{4}$

8.

$\frac{1}{2}$ $\frac{1}{4}$

9.

$\frac{1}{6}$ $\frac{1}{8}$

Complete the chart. Write a fraction for each part.

Fraction Model	Part that is Green	Part that is *not* Green
10. 	□/□	□/□
11.	□/□	□/□

Shade each set to represent the fraction.

12. $\frac{3}{6}$ shaded

○ ○ ○
○ ○ ○

13. $\frac{7}{8}$ shaded

○ ○ ○ ○
○ ○ ○ ○

 Problem Solving

14. Sydney has 8 packages of sticky notes. Five of the packages are pink, one is green, and two are blue. Write a fraction to show what part of the set of sticky notes is *not* pink.

□/□

Test Practice

15. What fraction of the figure is shaded?

Ⓐ $\frac{1}{2}$ Ⓒ $\frac{5}{8}$

Ⓑ $\frac{5}{6}$ Ⓓ $\frac{3}{8}$

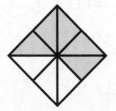

Name ....

Hands On
Fractions on a Number Line

Lesson 5

ESSENTIAL QUESTION
How can fractions
be used to represent
numbers and their parts?

Build It

Represent the fractions $\frac{1}{3}$, $\frac{2}{3}$, and $\frac{3}{3}$ on a number line.

1. Start at the left. Place a $\frac{1}{3}$-fraction tile
   above the number line. Trace the tile.
   Shade in the figure.

   What fraction does the shaded area represent? $\frac{\boxed{}}{\boxed{}}$

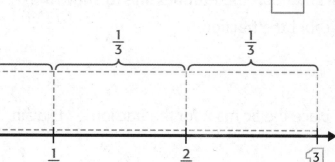

one whole

2. Place a second and third $\frac{1}{3}$-tile above the number line next to
   the shaded figure. Trace the tiles. Shade in the figures.

   What fraction do the first two shaded areas represent together? $\frac{\boxed{}}{\boxed{}}$

   What fraction do all three shaded areas represent? $\frac{\boxed{}}{\boxed{}}$

   How many parts does each fraction represent?

   $\frac{1}{3} =$ _____ part $\frac{2}{3} =$ _____ parts $\frac{3}{3} =$ _____ parts

Online Content at 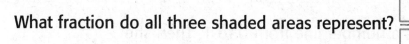 connectED.mcgraw-hill.com

Try It

Locate the fractions $\frac{1}{3}$ and $\frac{2}{3}$ on the number line.

1 One endpoint of the fraction tile that represents $\frac{1}{3}$ is at 0. Make a tic mark on the number line to show the other endpoint. Label the fraction.

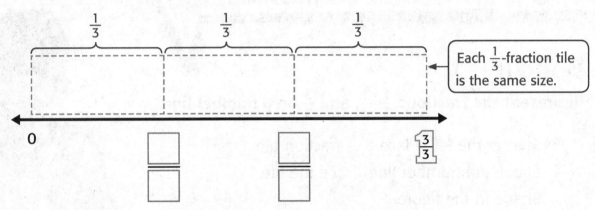

Each $\frac{1}{3}$-fraction tile is the same size.

2 One endpoint of the two fraction tiles that represent $\frac{2}{3}$ is at 0. Make a tic mark on the number line to show the other endpoint. Label the fraction.

Talk About It

1. Where would you place the tic mark for the fraction $\frac{3}{3}$? Explain.

_____

_____

2. Mathematical **PRACTICE** **6** **Be Precise** Sam is making a number line and wants to mark a point for $\frac{3}{4}$ on the number line. Into how many parts should he divide the number line? Explain.

_____

_____

3. Suppose Lisa makes a number line from 0 to 1. Then she divides the number line into 8 parts. What seven fractions will she place at each tic mark between 0 and 1?

_____

Practice It

Label each unknown with the fraction of the whole it represents.

4.

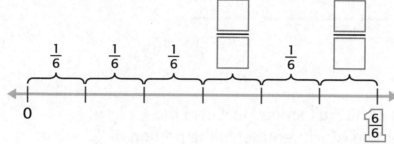

5.

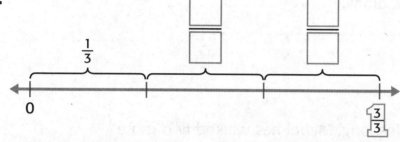

Label the fractions on the number lines between 0 and 1.

6.

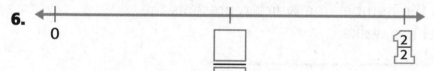

7.

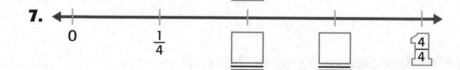

8.

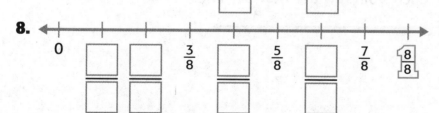

Write the point that represents each fraction.

9. $\frac{2}{6}$ is represented by point _____.

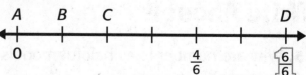

10. $\frac{4}{4}$ is represented by point _____.

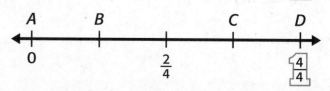

11. There are 6 students in science lab. Four of the students are girls. Label the fraction on the number line which represents the number of students that are girls.

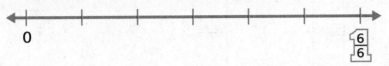

12. Mac drank two-fourths of his fruit smoothie. Label the fraction on the number line which represents the portion of the fruit smoothie Mac drank.

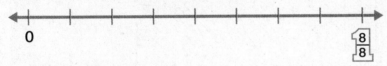

13. A walking trail is 8 miles long. Mabel has walked two more miles than Sherri. Sherri has walked three-eighths of the trail. Label the fraction on the number line which represents the part of the trail Mabel has walked.

Mathematical
14. **PRACTICE** 2 **Use Number Sense** Partition the number line into sixths. Label each point on the number line.

Write About It

15. Why are number lines helpful models to use to represent fractions?

_____

_____

MY Homework

Lesson 5

Hands On:
Fractions on a
Number Line

Homework Helper

Need help? connectED.mcgraw-hill.com

Hailey bought 6 apples. One apple is green and the rest are red. Label the fraction on the number line which represents the part of the apples that are red.

There are a total of 6 apples. So, the number line is divided into 6 parts.

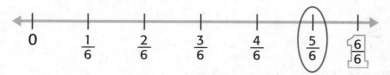

If 1 out of 6 apples is green, and the rest are red, then there are

5 red apples. The fraction that represents the red apples is $\frac{5}{6}$.

Practice

Label each unknown with the fraction of the whole it represents.

1.

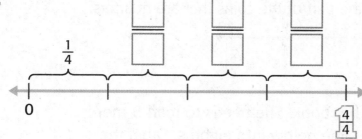

2.

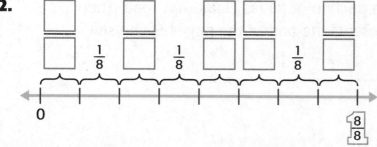

Label the fractions between 0 and 1 on the number lines.

3.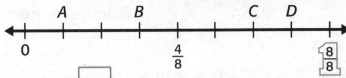

4.

Write the fraction that is represented by each point.

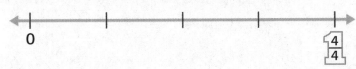

5. Point *A* is ☐/☐.

6. Point *D* is ☐/☐.

7. Point *C* is ☐/☐.

8. Point *B* is ☐/☐.

 Problem Solving

9. **Mathematical PRACTICE** ⑤ **Use Math Tools** Jamal spent $\frac{3}{4}$ of his allowance. On the number line, label the fraction that represents the part that Jamal spent.

10. Eric and Rachel have a total of 6 coins. One coin is a nickel, one is a dime, and the rest are quarters. On the number line, label the fraction that represents the part of the coins that are quarters.

11. Elaine has read 3 pages in her book. She needs to read 5 more pages. Partition the number line below into eighths. Label the fraction that represents the fraction of pages Elaine has read. Then label the fraction that represents the part of the pages Elaine still needs to read.

Name
....

Equivalent Fractions

Lesson 6

ESSENTIAL QUESTION
How can fractions
be used to represent
numbers and their parts?

Fractions that name the same part of the whole
are **equivalent fractions.** Equivalent fractions
are equal and have the same size.

 Math in My World Tools Watch Tutor

YUM!

Example 1

**Noah has one half of a fruit pizza. How could
Noah cut the pizza into smaller equal-sized
pieces and still have half of the whole pizza?**

1 Use $\frac{1}{4}$-fraction tiles to equal the length
of the $\frac{1}{2}$-fraction tile model.

How many $\frac{1}{4}$-tiles are there? _____

So, $\frac{1}{2} = \frac{2}{4}$. They are _____.
fractions. They represent the same part of one whole.

1	
$\frac{1}{2}$	$\frac{1}{2}$
$\frac{1}{4}$	$\frac{1}{4}$	$\frac{1}{4}$	$\frac{1}{4}$

2 Noah could cut half of the pizza into even
smaller equal-sized pieces.

| $\frac{1}{8}$ | $\frac{1}{8}$ | $\frac{1}{8}$ | $\frac{1}{8}$ | $\frac{1}{8}$ | $\frac{1}{8}$ | $\frac{1}{8}$ | $\frac{1}{8}$ |

Use $\frac{1}{8}$-fraction tiles to equal the length of the $\frac{1}{2}$-fraction
tile model.

How many $\frac{1}{8}$-tiles are there? _____

So, $\frac{1}{2} = \frac{2}{4} = \frac{4}{8}$. They are _____ fractions.

Half of the fruit pizza is the same as $\frac{2}{4}$ or $\frac{4}{8}$ of the pizza.

Online Content at connectED.mcgraw-hill.com

Example 2

Kwan's bookshelf has 3 shelves. Only one of the shelves has books. Kwan says $\frac{1}{3}$ of the shelves have books. His mom says $\frac{2}{6}$ of the shelves have books. Are they both correct?

Use fraction tiles and number lines to find if the fractions are equivalent.

One Way Use fraction tiles.

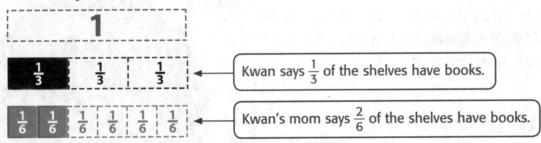

Kwan says $\frac{1}{3}$ of the shelves have books.

Kwan's mom says $\frac{2}{6}$ of the shelves have books.

One-third of one whole is equivalent to $\frac{2}{6}$.

Another Way Use number lines.

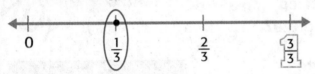

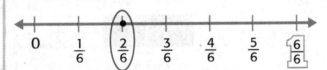

One-third of one whole shares the same point on the number line as $\frac{2}{6}$.

They are both correct. $\frac{1}{3} = \frac{2}{6}$ so, they are _____ fractions.

Talk MATH

What pattern do you see in the equivalent fractions $\frac{1}{2}$, $\frac{2}{4}$, $\frac{4}{8}$?

I Love Traveling!

Guided Practice ✓ Check

1. Complete the number sentence with an equivalent fraction.

$$\frac{1}{2} = \frac{\square}{4}$$

Independent Practice

Complete each number sentence to show equivalent fractions.

2.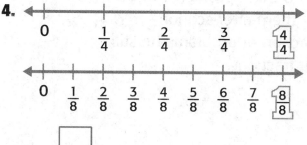

$$\frac{1}{2} = \frac{\boxed{}}{6}$$

3.

$$\frac{\boxed{}}{4} = \frac{6}{8}$$

4.

$$\frac{2}{4} = \frac{\boxed{}}{\boxed{}}$$

5.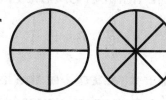

$$\frac{1}{3} = \frac{\boxed{}}{\boxed{}}$$

Match each pair of equivalent fractions.

6.

7.

8.

9.

Problem Solving

10. Raheem was at bat 8 times during a baseball game. He struck out 2 times. Circle the equivalent fractions that represent the number of times Raheem hit the ball.

$$\frac{2}{8} \qquad \frac{6}{8} \qquad \frac{1}{4} \qquad \frac{3}{4}$$

11. Leah said that it rained 2 of the last 4 school days. Circle the equivalent fractions that represent the number of days it rained over the last 4 school days.

$$\frac{2}{4} \qquad \frac{1}{2} \qquad \frac{4}{2} \qquad \frac{3}{4}$$

12. **Mathematical PRACTICE** ➋ **Reason** Andrew had 8 math problems to do for homework. He completed half of them after school. Shade the bar below to show the fraction of problems he still has to do. Then write two equivalent fractions.

HOT Problems

13. **Mathematical PRACTICE** ➌ **Find the Error** Circle the fraction that does not belong with the other 3. Explain.

$$\frac{4}{8} \qquad \frac{1}{2} \qquad \frac{8}{16} \qquad \frac{3}{4}$$

14. **? Building on the Essential Question** How do you know if two fractions are equivalent?

Name ....

MY Homework

Lesson 6

Equivalent Fractions

Homework Helper

Need help? connectED.mcgraw-hill.com

Marley packed 2 of the 4 apricots her mom just bought for her lunch. Find an equivalent fraction to represent the part of the apricots that Marley just packed.

1 **Represent the fraction on a number line.**
Divide a number line into four equal parts. Mark the fraction.

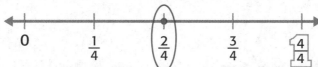

2 **Find an equivalent fraction.**
Draw another number line of equal length. Equally divide this number line another way. $\frac{2}{4}$ and $\frac{1}{2}$ name the same point.

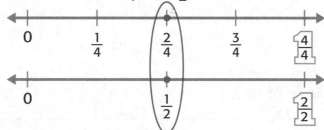

The number lines show that $\frac{2}{4}$ names the same point as $\frac{1}{2}$.

So, $\frac{2}{4}$ and $\frac{1}{2}$ are equivalent fractions.

Practice

Complete each number sentence to show equivalent fractions.

1.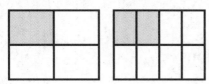

$$\frac{1}{4} = \frac{\boxed{}}{8}$$

2.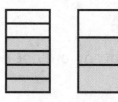

$$\frac{\boxed{}}{6} = \frac{\boxed{}}{3}$$

Complete each number sentence to show equivalent fractions.

3.

$$\frac{1}{\boxed{}} = \frac{3}{\boxed{}}$$

4.

$$\frac{\boxed{}}{\boxed{}} = \frac{\boxed{}}{\boxed{}}$$

 ## Problem Solving

5. Hiroshi made 6 puppets. Two of the puppets were dogs, two were cats, and two were birds. Circle the equivalent fractions that represent the part of the puppets that were cats.

$$\frac{1}{2} \qquad \frac{1}{3} \qquad \frac{2}{4} \qquad \frac{2}{6}$$

Mathematical
6. **PRACTICE** 2 **Use Number Sense** A rosebush had 8 blossoms. Two of the blossoms withered and fell off. Circle the equivalent fractions which represent the part of the blossoms still on the bush.

$$\frac{2}{8} \qquad \frac{7}{8} \qquad \frac{3}{4} \qquad \frac{6}{8}$$

Vocabulary Check

7. Write a definition for equivalent fractions. Then give an example.

_____

_____

Test Practice

8. Which of the following are *not* equivalent fractions?

Ⓐ $\frac{2}{6}$ and $\frac{1}{3}$ Ⓒ $\frac{1}{4}$ and $\frac{2}{8}$

Ⓑ $\frac{2}{3}$ and $\frac{4}{6}$ Ⓓ $\frac{1}{2}$ and $\frac{3}{8}$

Name _____

Number and Operations – Fractions

3.NF.1, 3.NF.2, 3.NF.2b, 3.NF.3, 3.NF.3a, 3.NF.3b, 3.NF.3c

CCSS

Fractions as One Whole

Lesson 7

ESSENTIAL QUESTION
How can fractions be used to represent numbers and their parts?

Whole numbers can be written as fractions. When the numerator is the same as the denominator, the fraction equals 1.

 Math in My World Tools Tutor

Example 1

There are 4 panes of glass in one window of Landon's classroom. Each pane of glass is $\frac{1}{4}$ of the whole window.

How many fourths equal 1 whole?

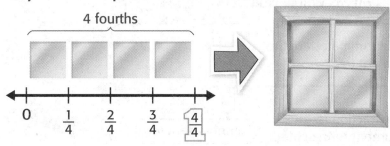

4 fourths

$$0 \quad \frac{1}{4} \quad \frac{2}{4} \quad \frac{3}{4} \quad \frac{4}{4}$$

_____ **fourths** = _____ **whole**

Write the fraction.

[ ] ← four parts

4 ← One whole is partitioned into four parts.

Place a point on the number line to graph this fraction. The models show that $\frac{4}{4}$ and 1 share the same point and have the same size.

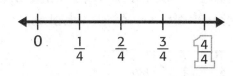

$$0 \quad \frac{1}{4} \quad \frac{2}{4} \quad \frac{3}{4} \quad \frac{4}{4}$$

So, 4 fourths = _____ whole. Or, $\frac{\Box}{\Box}$ = 1.

The fraction $\frac{1}{1}$ means 1 whole partitioned into 1 group. So, $\frac{1}{1} = 1$.

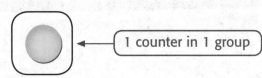

1	or

1 counter in 1 group

Example 2

How many wholes are in $\frac{3}{1}$?

Think: $\frac{3}{1}$ ← 3 parts

$\frac{3}{1}$ ← 1 whole partitioned into 1 part → ① 1

    1 part 2 parts 3 parts

So, ①①① $= \frac{3}{1}$ or _____ . There are _____ wholes in $\frac{3}{1}$.

Key Concept   Fractions as Whole Numbers

If the numerator and the denominator are the same, the fraction is equivalent to 1.

Example: $\frac{3}{3} = 1$ ←

If the denominator is 1, the fraction is equivalent to the whole number represented by the numerator.

number sentence

Example: $\frac{3}{1} = 3$ ←

Talk MATH

How can you tell whether $\frac{6}{1}$ is greater or less than 1?

Guided Practice

Write a fraction to represent the shaded part of each whole or set of wholes.

1.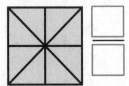

2.

Name
....

Independent Practice

Write a fraction to represent the shaded part of each whole or set of wholes.

3. 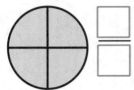   ⬜/⬜

4. 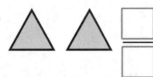   ⬜/⬜

5.   ⬜/⬜

6.    ⬜/⬜

7. 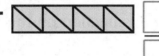   ⬜/⬜

8. 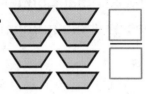   ⬜/⬜

Write each whole number as a fraction.

9. 4 = ⬜/⬜

10. 2 = ⬜/⬜

11. 6 = ⬜/⬜

12. 1 = ⬜/⬜

13. 8 = ⬜/⬜

14. 3 = ⬜/⬜

Find the missing numerators and denominators. Then circle the model that is not a fraction for 1 whole.

15.   ⬜/6

16.   ⬜/4

17.   ⬜/⬜

18.   2/⬜

19. Write three different fractions equivalent to 1.  ⬜/⬜ , ⬜/⬜ , ⬜/⬜

Problem Solving

20. Tim has 7 magazines. He gave all of them to Mike. Write a fraction that represents the part of the magazines Tim gave to Mike. Then write the fraction as a whole number.

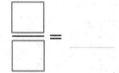 = _____

21. Conner has 3 cups of raisins. Write the number of cups of raisins that Conner has as a fraction. Then write the fraction as a whole number. Graph this fraction on the number line.

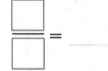 = _____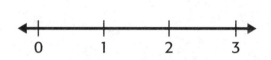

22. Mathematical **PRACTICE** ➊ **Make Sense of Problems** Carla took two photographs at a zoo. Two of the photographs were of giraffes. Write a fraction that represents the part of the photographs that were giraffes. Then write the fraction as a whole number.

 = _____

23. Mathematical **PRACTICE** ➍ **Model Math** Draw figures to represent $\frac{4}{1}$ and $\frac{6}{6}$ as whole numbers.

My Drawing!

24. ❓ **Building on the Essential Question** How can whole numbers be represented as fractions?

_____

_____

_____

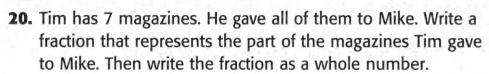

MY Homework

Lesson 7

Fractions as One Whole

Homework Helper

Need help? connectED.mcgraw-hill.com

How many sixths equal 1 whole? Write the fraction.

The number line shows one whole partitioned into six equal parts. Six $\frac{1}{6}$-fraction tiles are placed above the number line.

$\frac{1}{6}$ $\frac{1}{6}$ $\frac{1}{6}$ $\frac{1}{6}$ $\frac{1}{6}$ $\frac{1}{6}$

0    $\frac{1}{6}$ $\frac{2}{6}$ $\frac{3}{6}$ $\frac{4}{6}$ $\frac{5}{6}$ $\frac{6}{6}$

Each $\frac{1}{6}$-fraction tile represents *one-sixth.*

So, *six-sixths* is equivalent to one whole.

Write the fraction.

$\frac{6}{6}$ ← six parts

← One whole is partitioned into six parts.

Practice

Write a fraction to represent the shaded part of each whole.

1.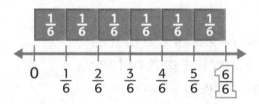

2.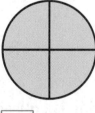

Write a fraction to represent each set of wholes.

3.

4.

Write each whole number as a fraction.

5. $8 = \dfrac{\square}{\square}$

6. $4 = \dfrac{\square}{\square}$

7. $2 = \dfrac{\square}{\square}$

8. $6 = \dfrac{\square}{\square}$

 Problem Solving

9. Gary sliced an apple into eighths. He gave eight of the pieces to his friends. Write a fraction that represents the part of the apple that was given to his friends. Then write this fraction as a whole number. Graph the fraction on the number line.

$$\dfrac{\square}{\square} = \underline{\quad}$$

Number line: $0 \quad \dfrac{1}{8} \quad \dfrac{2}{8} \quad \dfrac{3}{8} \quad \dfrac{4}{8} \quad \dfrac{5}{8} \quad \dfrac{6}{8} \quad \dfrac{7}{8} \quad \dfrac{8}{8}$

10. **Mathematical PRACTICE 2 Use Number Sense** The art teacher partitioned a piece of poster paper into three equal pieces. Each part was decorated for a school dance. Write a fraction that represents the part of the poster paper that was decorated for the school dance. Then write this fraction as a whole number.

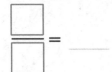

$$\dfrac{\square}{\square} = \underline{\quad}$$

Test Practice

11. Which of the following is equivalent to $\dfrac{4}{4}$?

Number line: $0 \quad \dfrac{1}{4} \quad \dfrac{2}{4} \quad \dfrac{3}{4} \quad \dfrac{4}{4} = ?$

Ⓐ $\dfrac{1}{4}$

Ⓒ $\dfrac{4}{1}$

Ⓑ 1

Ⓓ 4

Name _____

Number and Operations – Fractions
3.NF.1, 3.NF.2, 3.NF.2a, 3.NF.2b,
3.NF.3, 3.NF.3d

CCSS

Compare Fractions

Lesson 8

ESSENTIAL QUESTION
How can fractions be used to represent numbers and their parts?

You can compare fractions when the two fractions refer to the same size whole.

 Math in My World [Tools] [Watch] [Tutor]

Example 1

Ajay has $\frac{3}{6}$ of his homework done. Yuki finished $\frac{4}{6}$ of hers. Who has finished a greater part of their homework?

One Way Use fraction tiles.

Compare $\frac{3}{6}$ and $\frac{4}{6}$.

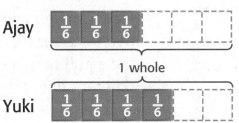

Ajay | $\frac{1}{6}$ $\frac{1}{6}$ $\frac{1}{6}$

1 whole

Yuki | $\frac{1}{6}$ $\frac{1}{6}$ $\frac{1}{6}$ $\frac{1}{6}$

The models show that $\frac{4}{6}$ has a greater size than $\frac{3}{6}$.

So, $\frac{4}{6} > \frac{3}{6}$.

Another Way Use a number line.

Represent each fraction on a number line.

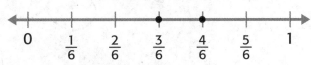

$$0 \quad \frac{1}{6} \quad \frac{2}{6} \quad \frac{3}{6} \quad \frac{4}{6} \quad \frac{5}{6} \quad 1$$

$\frac{4}{6}$ is closer to 1 whole. So, $\frac{4}{6} > \frac{3}{6}$.

So, _____ finished a greater part of her homework.

Online Content at [] connectED.mcgraw-hill.com

Example 2

Camille and Peter were reading a book. Camille read $\frac{1}{4}$ of the book, while Peter read $\frac{1}{3}$ of the same book. Who read a greater part of the book?

Compare $\frac{1}{4}$ and $\frac{1}{3}$.

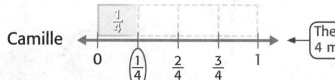

Camille

> The larger denominator of 4 means more, but smaller parts.

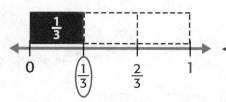

Peter

> The smaller denominator of 3 means fewer, but larger parts.

One-third is closer to 1 whole.

So, $\dfrac{\square}{\square}$ is greater than $\frac{1}{4}$ or $\dfrac{\square}{\square} > \frac{1}{4}$.

So, _____ read a greater part of the book.

Talk MATH

How can you compare two fractions that have the same numerator but different denominators?

Guided Practice ✓ Check

1. Use the models to compare. Use >, <, or =.

$$\frac{1}{3} \bigcirc \frac{2}{3}$$

Independent Practice

Use the models to compare. Use >, <, or =.

2.

| 0 | $\frac{1}{4}$ | $\frac{2}{4}$ | $\frac{3}{4}$ | $\frac{4}{4}$ |

| 0 | $\frac{1}{6}$ | $\frac{2}{6}$ | $\frac{3}{6}$ | $\frac{4}{6}$ | $\frac{5}{6}$ | $\frac{6}{6}$ |

$\frac{1}{4}$ ◯ $\frac{1}{6}$

3.

| 0 | $\frac{1}{8}$ | $\frac{2}{8}$ | $\frac{3}{8}$ | $\frac{4}{8}$ | $\frac{5}{8}$ | $\frac{6}{8}$ | $\frac{7}{8}$ | $\frac{8}{8}$ |

| 0 | $\frac{1}{8}$ | $\frac{2}{8}$ | $\frac{3}{8}$ | $\frac{4}{8}$ | $\frac{5}{8}$ | $\frac{6}{8}$ | $\frac{7}{8}$ | $\frac{8}{8}$ |

$\frac{7}{8}$ ◯ $\frac{6}{8}$

4.

$\frac{2}{4}$ ◯ $\frac{4}{8}$

5.

$\frac{4}{8}$ ◯ $\frac{4}{6}$

Use the number line for Exercises 6–8.

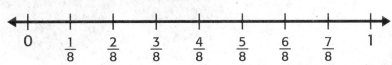

| 0 | $\frac{1}{8}$ | $\frac{2}{8}$ | $\frac{3}{8}$ | $\frac{4}{8}$ | $\frac{5}{8}$ | $\frac{6}{8}$ | $\frac{7}{8}$ | 1 |

6. Circle all of the fractions on the number line that are greater than $\frac{5}{8}$.

7. Draw a box around all of the fractions on the number line that are less than $\frac{3}{8}$.

8. Write the fraction from the number line that is greater than $\frac{3}{8}$ but less than $\frac{5}{8}$. $\frac{\boxed{}}{\boxed{}}$

Problem Solving

9. **Mathematical PRACTICE** 6 **Explain to a Friend** Alister makes a party mix with $\frac{1}{3}$ cup of raisins and $\frac{2}{3}$ cup of cereal. Are there more raisins or cereal? Explain.

_____

_____

_____

10. Debbie realized that more than $\frac{4}{8}$ of her summer vacation has passed. Circle the fraction that is greater than $\frac{4}{8}$.

$$\frac{5}{8} \qquad \frac{3}{8} \qquad \frac{2}{8}$$

HOT Problems

11. **Mathematical PRACTICE** 3 **Justify Conclusions** Is $\frac{1}{4}$ of the smaller waffle equal to $\frac{1}{4}$ of the larger waffle? Explain.

_____

_____

12. If you are comparing the fractions $\frac{4}{8}$ and $\frac{3}{8}$, how can you tell which fraction is greater without using models?

_____

_____

13. **Building on the Essential Question** How can fractions be compared?

_____

_____

_____

Name ....

MY Homework

Lesson 8

Compare Fractions

Homework Helper

Need help? connectED.mcgraw-hill.com

Luis and Malcolm are both on the tennis team. Luis has won $\frac{1}{2}$ of his matches. Malcolm has won $\frac{1}{6}$ of his matches. They both played the same number of matches. Who has won the greater fraction of matches?

Compare $\frac{1}{2}$ and $\frac{1}{6}$.

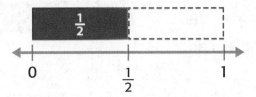

Luis

The smaller denominator of 2 means fewer, but larger parts.

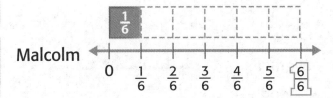

Malcolm

The larger denominator of 6 means more, but smaller parts.

The models show that $\frac{1}{2}$ has a greater size than $\frac{1}{6}$.

So, $\frac{1}{2} > \frac{1}{6}$. Luis won a greater fraction of matches.

Practice

Use the models to compare. Use >, <, or =.

1.

$\frac{1}{3} \bigcirc \frac{1}{2}$

2.

$\frac{5}{6} \bigcirc \frac{4}{6}$

Use the models to compare. Use >, <, or =.

3.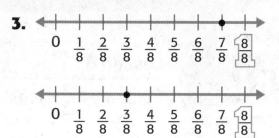

$\dfrac{7}{8} \bigcirc \dfrac{3}{8}$

4.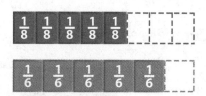

$\dfrac{2}{3} \bigcirc \dfrac{4}{6}$

Problem Solving

5. **Mathematical PRACTICE 3** **Justify Conclusions** Harvey practiced the piano for $\dfrac{5}{8}$ of an hour. Annika practiced the piano for $\dfrac{5}{6}$ of an hour. Use the models to determine who practiced the piano for a longer period of time.

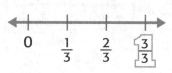

6. **Mathematical PRACTICE 5** **Use Math Tools** The average housecat sleeps about $\dfrac{2}{3}$ of a day. Most people sleep about $\dfrac{1}{3}$ of the day. Do housecats or people sleep for a greater fraction of the day? Graph both fractions on the number line to compare.

Test Practice

7. The number line shows which of the following fractions is less than $\dfrac{2}{4}$?

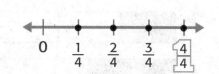

(A) $\dfrac{1}{4}$ (C) $\dfrac{3}{4}$

(B) $\dfrac{2}{4}$ (D) $\dfrac{4}{4}$

Review

Vocabulary Check

Draw lines to match each with its definition.

1. equivalent fractions

- Any fraction with a numerator of 1. Examples: $\frac{1}{2}$, $\frac{1}{3}$, and $\frac{1}{4}$.

2. denominator

- Fractions that have the same value. Example: $\frac{2}{4} = \frac{1}{2}$.

3. fraction

- The number above the bar in a fraction; the part of the fraction that tells how many of the equal parts are being represented.

4. numerator

- The number below the bar in a fraction; the part of the fraction that tells the total number of equal parts.

5. unit fraction

- A number that represents part of a whole or part of a set.

Concept Check

Write how many equal parts. Shade one part. Write its unit fraction.

6.

_____ equal parts

unit fraction: _____

7.

_____ equal parts

unit fraction: _____

Write the missing numerator or denominator to show the shaded part.

8. $\dfrac{3}{\square}$

9. $\dfrac{\square}{6}$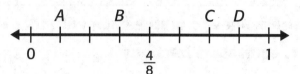

---

10. Write the point that represents the fraction.

$\dfrac{6}{8}$ is represented by point _____ .

Complete each number sentence to show equivalent fractions.

11.

$\frac{1}{4}$	$\frac{1}{4}$		

$\frac{1}{8}$	$\frac{1}{8}$	$\frac{1}{8}$	$\frac{1}{8}$				

$\dfrac{2}{4} = \dfrac{\square}{8}$

12.

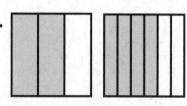

$\dfrac{2}{\square} = \dfrac{4}{\square}$

Write each whole number as a fraction.

13. $3 = \dfrac{\square}{\square}$

14. $6 = \dfrac{\square}{\square}$

15. $4 = \dfrac{\square}{\square}$

Use the models to compare. Use >, <, or =.

16.

$\frac{1}{8}$							

$\frac{1}{8}$	$\frac{1}{8}$	$\frac{1}{8}$					

$\dfrac{1}{8} \bigcirc \dfrac{3}{8}$

17.

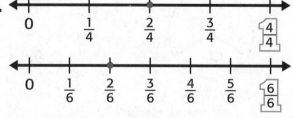

$\dfrac{2}{4} \bigcirc \dfrac{2}{6}$

Problem Solving

18. Emma has 3 black cats and 1 gray cat. What fraction of the set of cats is gray?

_____

19. Lucy walks $\frac{3}{4}$ of a mile. Sergio walks $\frac{3}{6}$ of a mile. Who walks farther?

_____

Solve Exercises 20 and 21 by drawing a diagram.

20. A music CD tower can hold 10 compact discs. One-half of the slots are filled with CDs. How many CDs are in the CD tower?

_____

21. There are 16 students in the school play. One-fourth of the students are wearing yellow costumes. Six students are wearing purple costumes. The remaining students are wearing orange costumes. How many students are wearing orange costumes?

_____

Test Practice

22. Melinda and Ruben are playing tic-tac-toe. Melinda has Xs in one-third of the 9 places. Ruben has Os in 2 of the places. How many squares are empty?

Ⓐ 3 squares   Ⓒ 7 squares

Ⓑ 4 squares   Ⓓ 11 squares

Reflect

Use what you learned about fractions to complete
the graphic organizer.

Part of One Whole

Part of a Set

**ESSENTIAL
QUESTION**

How can fractions be
used to represent
numbers and their parts?

Equivalent Fractions

Compare Fractions

Reflect on the ESSENTIAL QUESTION . Write your answer below.

_____

_____

ESSENTIAL QUESTION

Why do we measure?

Around My House

Watch a video!

Watch

MY Common Core State Standards

CCSS

Copyright © The McGraw-Hill Companies, Inc.   Mel Yates/Photodisc/PunchStock

Name

Check ✓ ← Go online to take the Readiness Quiz

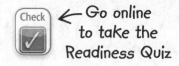

Circle the object that holds more.

1.

2.

3. Which should hold more water, a swimming pool or a bathtub?

_____

Circle the object which weighs more.

4.

5.

6. Consuela took two equal-sized glasses of milk to the table. One of the glasses was full, and the other was half full. Which glass had less milk?

_____

Write the time shown on each clock.

7.  _____

8.  _____

Shade the boxes to show the problems you answered correctly.

How Did I Do? → | 1 | 2 | 3 | 4 | 5 | 6 | 7 | 8 |

Online Content at ↗ connectED.mcgraw-hill.com

625

MY Math Words

Vocab
abc

Review Vocabulary

heavier   hour   lighter   minute   second

Making Connections

Sort the review vocabulary into the correct sections of the
diagram. Then answer the question.

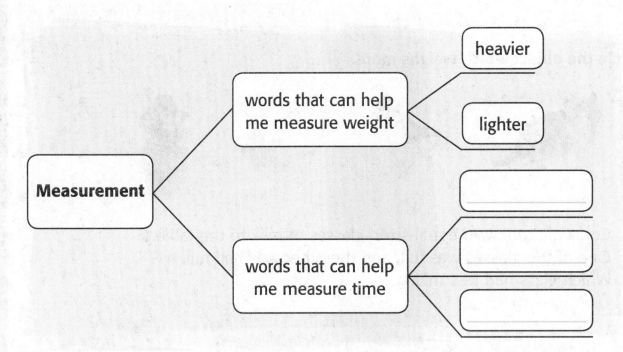

How did you decide where to place each of the review
vocabulary words?

_____

_____

_____

MY Vocabulary Cards

Lesson 11–5

analog clock

Lesson 11–1

capacity

Lesson 11–5

digital clock

Lesson 11–3

gram (g)

I gram

Lesson 11–3

kilogram (kg)

I kilogram

Lesson 11–1

liquid volume

Lesson 11–1

liter (L)

I liter   **5 liters**

Lesson 11–3

mass

less mass   **more mass**

Ideas for Use

- Identify a relationship between two or more words. Display the cards. Ask a friend to guess the relationship the cards represent.

- Draw or write examples for each card. Be sure your examples are different from what is shown on each card.

The amount of liquid a container can hold. Also known as *liquid volume*.

Describe a real-life example of when you might need to know a container's capacity.

_____

_____

A clock that has both hour and minute hands.

Explain how the hour hand and minute hand show a part-whole relationship.

_____

_____

A metric unit for measuring mass.

How can you remember that a gram has less mass than a kilogram?

_____

_____

A clock that uses only numbers to show time.

Describe one advantage of using a digital clock instead of an analog clock.

_____

_____

The amount of liquid a container can hold. Also known as *capacity.*

Volume is a multiple-meaning word. What does it mean in this sentence? *Please turn up the volume on the radio.*

_____

A metric unit for measuring mass.

Would you measure the mass of a large dog in grams or kilograms? Explain.

_____

_____

The amount of matter, or material, in an object.

Which has more mass, a horse or a kitten? Explain.

_____

A metric unit for measuring capacity.

What kind of containers might hold more than 1 liter of liquid?

_____

_____

MY Vocabulary Cards

Mathematical PRACTICE

Lesson 11–1

metric unit

I gram   I liter

Lesson 11–1

milliliter (mL)

I milliliter = 10 drops

Lesson 11–6

time interval

Start   **End**

Lesson 11–1

unit

I kilogram   I minute

Ideas for Use

- Ask students to arrange two cards to show a pair. Have them explain the meaning of their pairing.

- Use the blank cards to write words from a previous chapter that you would like to review.

- -

A metric unit for measuring capacity.

Hudson's hamster drinks about 10 milliliters of water each day. About how much water does Hudson's hamster drink in seven days?

_____

A unit of measure in the metric system.

Name two other metric units besides gram and liter that are included in this chapter.

_____

One specific amount of measurement.

Identify the two units of capacity used in Lesson 1.

_____

The time that passes from the start of an activity until the end of an activity.

The prefix *inter-* means "between." How does this help you remember the meaning of *time interval*?

_____

_____

MY Foldable

FOLDABLES® Follow the steps on the back to make your Foldable.

Mass

The amount of _____ , or stuff in an object.

A paper clip has a mass of about _____ gram.

A baseball bat has a mass of about _____ kilogram.

g

kg

Capacity

The amount of liquid a container can _____ .

A dropper holds about _____ milliliter.

A water bottle holds about _____ liter.

mL

L

 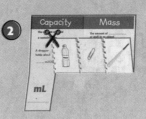

Measurement and Data
3.MD.2

CCSS

Hands On
Estimate and Measure Capacity

Lesson 1
ESSENTIAL QUESTION
Why do we measure?

The amount of liquid a container can hold is called its **capacity**. Capacity is also known as **liquid volume**.

One **unit** is one specific amount of measurement.
A **metric unit** is a unit of measure in the metric system. One metric unit of capacity is a **liter (L)**.

This water bottle holds about 1 liter of water.

Use liters to measure containers of greater capacity.

Measure It

How much liquid is a liter?

1. Find three containers. Estimate whether each holds less than, about, or more than 1 liter. Write the name of the container. Mark your estimate. An example is shown.

	My Estimate			My Measure
Container	Less	About	More	Actual
water glass		X		

2. Pour liquid from the container into the 1-liter measuring cup to check each estimate. Record your results in the table.

To measure the liquid capacity of smaller containers, you need a smaller unit of measure.

A smaller metric unit of capacity is a **milliliter (mL)**.

A dropper holds about 1 milliliter of liquid.

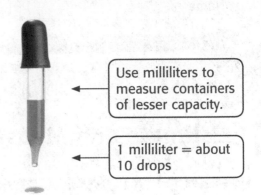

Use milliliters to measure containers of lesser capacity.

1 milliliter = about 10 drops

Try It

How much liquid is a milliliter?

1 About how many milliliters do you think a teaspoon holds? Estimate its capacity. Record your estimate in the table.

2 Fill a teaspoon with water. Empty it into a metric measuring cup. Continue until the liquid is up to the 10 mL marking.

About how many teaspoons did you use?

about _____ teaspoons

Together, about _____ teaspoons have the capacity of 10 mL. What is the capacity of

1 teaspoon then? _____

Container	My Estimate (mL)	Actual (mL)
teaspoon		
paper cup		

| 100 mL
| 80
| 60
| 40
| 20

3 Estimate the capacity of a paper cup, and two other small containers. Record each estimate in the table.

4 Pour liquid from each small container into the measuring cup to check each estimate. Record your results in the table.

Talk About It

1. **Mathematical PRACTICE ❸ Justify Conclusions** How did you determine an estimate for the paper cup?

_____

_____

Name

Practice It

Circle the better unit to measure each capacity.

2. pot of soup

milliliter   liter

3. juice box

milliliter   liter

4. wading pool

milliliter   liter

5. bottle of glue

milliliter   liter

6. pitcher of lemonade

milliliter   liter

7. fish aquarium

milliliter   liter

Circle the better estimate.

8.

400 mL   400 L

9.

4 mL   40 L

10.

10 mL   10 L

11.

900 mL   900 L

12.

2 mL   2 L

13.

100 mL   100 L

How much water is in each container? Circle the amount.

14.

500 mL
400
300
200
100

100 mL

300 mL

15.

500 mL
400
300
200
100

200 mL

500 mL

Apply It

16. Riley bought 2 juice boxes. The capacity of one juice box is shown. What is the total capacity of 2 juice boxes?

FRUIT PUNCH 50 mL

17. **Mathematical PRACTICE** 6 **Be Precise** Avery's recipe asks for 100 mL of milk. She only has a 10 mL measuring cup. How many times will she need to fill her measuring cup to get the right amount of milk for her recipe? Explain.

18. Anna recorded the amount of water she gave her tomato plants for the past 5 days. The table shows the amounts of water Anna recorded. How much water did Anna give her tomato plants the last 5 days?

Record of This Week's Water				
14 mL	12 mL	4 mL	26 mL	10 mL

19. **Mathematical PRACTICE** 4 **Model Math** Name 3 items sold in a grocery store that are packaged in liter containers.

Rain, Rain, Go Away!

Write About It

20. Why are there two different units of measure for measuring capacity?

MY Homework

Homework Helper

Need help? connectED.mcgraw-hill.com

Is 1 liter or 1 milliliter a more reasonable estimate for a drop of food coloring Brianne will use to color icing for baking?

Use a liter to measure greater capacity.

Use a milliliter to measure lesser capacity.

A drop of food coloring is a small amount. It is unreasonable to estimate that one drop is a liter of liquid.

So, 1 milliliter of food coloring is a more reasonable estimate.

Practice

Circle the better unit to measure each capacity.

1.

    milliliter   liter

2.

    milliliter   liter

3.

    milliliter   liter

4.

    milliliter   liter

How much water is in each container? Circle the amount.

5.  200 mL

400 mL

6.  100 mL

200 mL

 Problem Solving

7. **Mathematical** **PRACTICE 5** **Use Mental Math** Ian is going to water his plants. Is it reasonable to say he will fill the watering can with 3 liters of water? Explain.

_____

_____

8. Gianna is heating a can of soup for lunch. Is it reasonable to say she is heating 3 milliliters or 300 milliliters of soup? Explain.

_____

_____

9. Vincent is painting all four walls of his bedroom. Is it more reasonable to measure the paint he will use in milliliters or liters? Explain.

_____

_____

Vocabulary Check [Vocab]

Match each vocabulary word or set of words to its meaning.

10. capacity/liquid volume

11. liter (L)

12. milliliter (mL)

13. unit

14. metric units

• a metric unit used for greater capacities

• a standard for measurement

• the amount of liquid a container can hold

• a metric unit used for smaller capacities

• a unit of measure in the metric system.

Name _____

Solve Capacity Problems

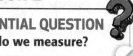

 ## Math in My World

Example 1

Emily used 240 milliliters of lemon juice and 960 milliliters of water to make lemonade. How many milliliters of lemonade will she make?

Add to find the unknown.

$240 + 960 = ?$

$240 + 960 =$ _____ milliliters

The unknown is _____ milliliters.

Emily will make _____ milliliters of lemonade.

Example 2

The total capacity of 8 pitchers is 24 liters. What is the capacity of each pitcher if each has an equal amount of lemonade? Write an equation with a symbol for the unknown. Then solve.

Find the unknown. $24 \div 8 = \blacksquare$

There are _____ liters in each of the pitchers.

The unknown is _____ .

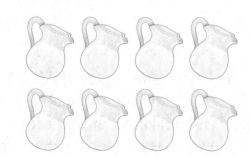

Check

Use the inverse operation to check.

$8 \times 3 = 24$ The answer is reasonable.

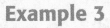

Example 3

Dylan is helping his dad wash the car. His dad filled up 2 buckets with soapy water and 2 buckets with clean water. The capacity of each bucket is 9 liters. What is the total capacity of the 4 buckets?

Find 4 × 9. Label the capacity of each bucket.

4 × 9 = _____   There are _____ liters of water in the buckets.

Guided Practice

Algebra Write an equation with a symbol for the unknown. Then solve.

1. Find the total capacity of the liquid shown in the containers below.

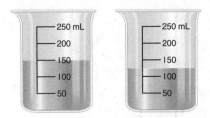

2. Peyton's tea kettle holds 2 liters, or 2,000 milliliters of water. She uses 350 milliliters of water for a cup of tea. How much water is left in the kettle?

Talk MATH

Look back at Exercise 2. How did you know what operation to use?

Independent Practice

Algebra Write an equation with a symbol for the unknown. Then solve.

3. Find the total capacity of the liquid shown in the containers below.

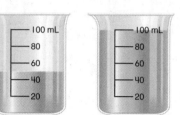

4. How much liquid would be left if you poured out 250 milliliters?

5. If you equally pour this liquid into three containers how much liquid would be in each container?

6. The capacity of 1 pitcher is shown. What is the total capacity in liters, of 2 pitchers?

7. One small carton of milk is 250 milliliters. One large carton of milk is 1,000 milliliters, or 1 liter. Circle the correct number of small milk cartons equal to 1,000 milliliters, or 1 liter.

250 + _____ + _____ + _____ = 1,000 milliliters or 1 L

Problem Solving

8. Carlos' aquarium holds 81 liters of water. His pail holds 9 liters of water. How many pails of water does it take to fill the aquarium?

9. There are 600 milliliters of water in Lisa's vase. If she pours out 254 milliliters of water, how much water will be left in the vase?

10. Tara has 3 coolers with 7 liters of lemonade in each. How many liters of lemonade does she have in all?

11. **Mathematical PRACTICE** **5** **Use Mental Math** Sean's mother bought 2 bottles of shampoo. Each bottle contains 800 milliliters of shampoo. How many milliliters of shampoo did Sean's mother buy?

HOT Problems

12. **Mathematical PRACTICE** **1** **Plan Your Solution** A cooking pot holds 1,000 milliliters of sauce. Then half of the sauce is used. Another 300 milliliters of sauce is added back in. Once again, half of the pot of sauce is used. How many milliliters of sauce is needed in order to fill the pot again?

13. ❓ **Building on the Essential Question** Explain how you might decide whether to measure capacity in milliliters or liters.

MY Homework

Homework Helper

Need help? connectED.mcgraw-hill.com

Phillippe is conducting a science experiment. He puts 2 milliliters of liquid in each of 7 test tubes. How much liquid does Phillippe use altogether?

1 Write an equation with a symbol for the unknown.

$2 \times 7 = \blacksquare$

2 Multiply to find the answer.

$2 \times 7 = 14$

So, Phillippe used 14 milliliters of liquid altogether.

Practice

Algebra Write an equation with a symbol for the unknown. Then solve.

1. How much liquid would there be if you poured another 35 mL into the container?

2. If you poured this liquid in equal amounts into 10 separate containers, how much liquid would be in each container?

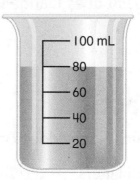

Problem Solving

3. Each stew pot can hold 6 liters of stew. If there are 5 stew pots, how much stew could be made at one time?

Mathematical
4. PRACTICE 6 **Be Precise** Kathleen needs 550 milliliters of cream for a recipe. Does Kathleen have enough cream for her recipe? Explain.

175 mL   350 mL

5. Glenna put 850 milliliters of gravy into a gravy boat. John poured 75 milliliters of gravy on his mashed potatoes. How much gravy is left in the gravy boat?

6. Mrs. Hudson made 25 liters of punch to equally pour into 5 punch bowls. How much punch will be poured into each bowl?

7. Claude has 24 milliliters of mouthwash left. He will use the same amount of mouthwash each day for the next three days. How much mouthwash does Claude use each day?

My Work!

Test Practice

8. How much honey does Omar use in five days?

Ⓐ 35 milliliters   Ⓒ 24 milliliters

Ⓑ 25 milliliters   Ⓓ 11 milliliters

Day	Honey
Monday	5 mL
Tuesday	5 mL
Wednesday	5 mL
Thursday	5 mL
Friday	5 mL

Hands On
Estimate and Measure Mass

Mass is the amount of matter, or material in an object. One metric unit of mass is the **kilogram (kg)**. These are items that have a mass of about 1 kilogram.

400 pennies   baseball bat   dictionary

> Use kilograms to measure objects of greater mass.

Measure It

What does 1 kilogram feel like?

1 Find three objects and estimate whether each feels less than, about, or more than 1 kilogram. Write the name of the object and your estimate. An example is shown.

	My Estimate			My Measure
Object	Less	About	More	Actual
baseball		X		

2 Place 8 rolls of pennies, on one side of the bucket balance. Check each estimate by placing one object on the other side of the balance. Record your results in the table.

Online Content at **connectED.mcgraw-hill.com**

A smaller unit of mass is the **gram (g)**.

Try It

What does 1 gram feel like?

These are items that have a mass of about 1 gram.

paper clip   base-ten cube

Use grams to measure objects of lesser mass.

1. Find three objects that you think would best be measured in grams. Write the name of the object and your estimate. An example is shown.

| | My Estimate | | | My Measure |
Object	Less	About	More	Actual
dollar bill		X		

2. Place a base-ten cube on one side of the bucket balance. Check each estimate by placing one object on the other side of the balance. Record your results in the table.

Talk About It

1. What characteristics were shared by the objects that you chose to measure the mass of in grams?

_____

_____

2. Mathematical PRACTICE 3 Draw a Conclusion Does a large object always have a greater mass than a small object? Explain.

_____

_____

Practice It

Circle the better unit to measure each mass.

3. toothbrush

gram   kilogram

4. television

gram   kilogram

5. shovel

gram   kilogram

6. teddy bear

gram   kilogram

7. pair of sunglasses

gram   kilogram

8. lawn mower

gram   kilogram

Circle the better estimate for each mass.

9.

5 g   5 kg

10.

50 g   5,000 g

11.

4 g   4 kg

12.

15 g   15 kg

13.

900 g   900 kg

14.

5 kg   5 g

Apply It

15. A store in the mall sells fresh baked pretzels. Would it be more reasonable to measure the mass of a baked pretzel in grams or kilograms? Explain.

_____

16. Mathematical **PRACTICE** 4 **Model Math** Mrs. Charles grows squash in her garden. Which would be a more reasonable estimate for the mass of a squash; 500 grams or 500 kilograms? Explain.

_____

_____

17. Mathematical **PRACTICE** 2 **Reason** A bag of potatoes has a mass of about 3 kilograms. Name two other items that have about the same mass. Explain your reasoning.

_____

_____

18. Mathematical **PRACTICE** 3 **Find the Error** Jack held an apple in his hand. He said, "This apple feels like it has a mass of 100 kilograms." Explain his error. What should Jack have said?

_____

_____

Write About It

19. What is different about the items you measured in grams and those you measured in kilograms?

_____

_____

MY Homework

Homework Helper   Need help? ⟋ connectED.mcgraw-hill.com

Gordon wants to measure the mass of his German shepherd. Is it more reasonable for Gordon to measure his dog's mass in grams or kilograms?

A paper clip has a mass of about 1 gram. A textbook has a mass of about 1 kilogram.

Because Gordon's dog is large, it does not make sense to measure its mass in grams. It is more reasonable for Gordon to measure his dog's mass in kilograms.

Practice

Circle the better unit to measure each mass.

1. bowling ball

   gram   kilogram

2. zebra

   gram   kilogram

3. cell phone

   gram   kilogram

4. laptop computer

   gram   kilogram

5. pair of socks

   gram   kilogram

6. a marble

   gram   kilogram

Circle the better estimate for each mass.

7.

4 grams   4 kilograms

8.

2 grams   2 kilograms

Circle the better estimate for each mass.

9.

20 grams   20 kilograms

10.

3 grams   3 kilograms

 ## Problem Solving

11. **Mathematical PRACTICE** **Use Mental Math** Dylan is making a casserole that serves 4 people. The recipe calls for shredded cheese. Would it be more reasonable for Dylan to measure the mass of the cheese in grams or kilograms? Explain.

_____

_____

_____

12. Paulette needs help moving her file cabinet. Is 40 grams or 40 kilograms a more reasonable estimate for the file cabinet's mass? Explain.

_____

_____

Vocabulary Check

Choose the correct word to complete each sentence.

mass   gram   kilogram

13. The amount of matter an object has is its _____.

14. A baseball bat has a mass of about 1 _____.

15. A penny has a mass of about 1 _____.

Measurement and Data
3.MD.2, 3.OA.3

CCSS

Solve Mass Problems

Lesson 4

ESSENTIAL QUESTION
Why do we measure?

 Math in My World Tools Tutor

Example 1

The bucket balance shows the mass of one marker. What is the mass of 5 markers?

Find the unknown. $8 \times 5 = \blacksquare$.

A bucket balance can be used to determine mass.

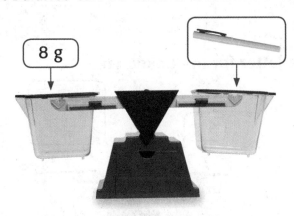

8 g

Mass Weights

1 Kg 500 g

200 g 100 g

50 g 20 g

10 g 5 g 1 g

Use mass weights on one side of the balance and the object on the other side. When the sides are balanced, the objects have the same mass.

Five markers are 8 grams each. So multiply.

$8 \times 5 =$ _____

The unknown is _____ grams.

So, the mass of 5 markers is _____ grams.

Helpful Hint
You can also use repeated addition.
$8 + 8 + 8 + 8 + 8 = 40$

Example 2

What is the total mass of the two pieces of fruit? Write an equation with a letter for the unknown. Then solve.

216 grams

132 grams

Add to find the unknown.

$216 + 132 = f$ ← unknown

$216 + 132 = $ _____ grams

The unknown is _____ .

So, the two pieces of fruit have a combined mass of 348 grams.

Guided Practice ✓ Check

Algebra Write an equation with a letter for the unknown. Then solve.

1. What is the mass of 9 buttons?

_____

 4 g

2. What is the mass of 1 jack?

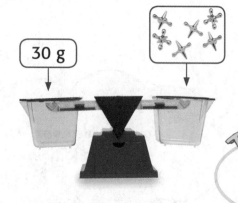 30 g

Talk MATH
Explain how you solved Exercise 2.

Independent Practice

Algebra **Write an equation with a letter for the unknown. Then solve.**

3. What is the mass of two of these books?

363 g

_____

4. The golf ball has a mass of 45 grams. What is the mass of the baseball?

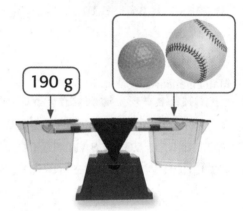
190 g

_____

5. What is the mass of one of these dominoes?

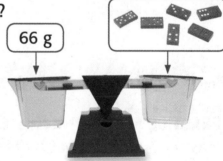
66 g

_____

Algebra **Find each unknown.**

6. ▪ kilograms × 9 = 63 kilograms   The unknown is _____.

7. 392 grams + ▪ grams = 523 grams   The unknown is _____.

8. 932 grams − ▪ grams = 149 grams   The unknown is _____.

9. 60 kilograms ÷ ▪ = 6 kilograms   The unknown is _____.

Problem Solving

10. A bucket balance shows that 2 rocks have a total mass of 732 grams. When 1 rock is removed, the bucket balance shows that the remaining rock is 428 grams. What is the mass of the rock that was removed?

My Work!

11. A student has a mass of 34 kilograms. An adult has a mass of 56 kilograms. What is the difference in mass between the student and adult?

Mathematical
12. PRACTICE 7 Identify Structure A penny has a mass of about 3 grams. About how many grams would one dollar's worth of pennies be? Write an equation with a letter for the unknown. Then solve.

HOT Problems

Mathematical
13. PRACTICE 6 Be Precise One button has a mass of 3 grams. What is the mass of one pencil?

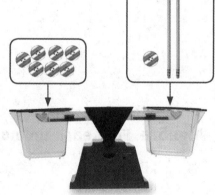

14. Building on the Essential Question When do you think it would be best to estimate the mass of an object and when would it be best to get an exact measure?

MY Homework

Homework Helper

Need help? connectED.mcgraw-hill.com

Three small packets of peanuts have a total mass of
30 grams. Each packet has the same mass. What
is the mass of 1 packet of peanuts?

1 Write an equation with
a letter for the unknown.

$30 \div 3 = m$

2 Divide to find the answer.

$30 \div 3 = 10$

$m = 10$ grams

So, 1 packet of peanuts has a mass of 10 grams.

Practice

Algebra Write an equation with a letter for the unknown.
Then solve.

1. What is the mass of 2 boxes of raisins?

_____

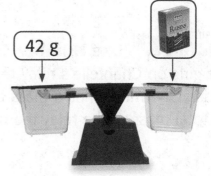

42 g

2. The scissors have a mass of 90 grams.
What is the mass of the stapler?

_____

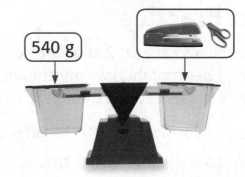

540 g

Algebra Solve for the unknown.

3. 511 kilograms + = 720 kilograms

    = _____ kilograms

4. 90 grams ÷ ■ = 10 grams

   ■ = _____

5. ■ grams − 138 grams = 704 grams

   ■ = _____

6. ■ × 20 kilograms = 80 kilograms

   ■ = _____

Problem Solving

7. Melinda's pair of ski boots and skis have a mass of 6 kilograms. What is the mass of one ski boot and one ski?

   _____

8. A full bag of oatmeal has a mass of 560 grams. Cooper uses 80 grams of oatmeal to make breakfast for his family. What is the mass of the remaining oatmeal?

   _____

9. Each bag can hold a mass of 3 kilograms. Wade has 24 kilograms of fruit to divide equally. How many bags will he need?

   _____

10. **Mathematical PRACTICE 2 Use Number Sense** Each of the four girls on Chantelle's relay team got a swimming medal. One medal has a mass of 16 grams. What is the total mass of the relay team's medals?

    _____

Test Practice

11. Jennifer's cats eat 2 kilograms of food per month. How much food do the cats eat in 8 months?

    Ⓐ 4 kilograms   © 14 kilograms

    ⓑ 10 kilograms   ⓓ 16 kilograms

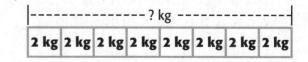

Check My Progress

Vocabulary Check

Label each with the correct vocabulary term.

capacity	**gram**	**kilogram**	**liter**
liquid volume	**mass**	**milliliter**	**unit**

1. One _____ is one specific amount of measurement.

2. _____ is the amount of matter, or material, in an object.

3. _____, or _____, is the amount a container can hold.

4. One metric unit of capacity is the _____, and is equal to 1,000 milliliters.

5. A metric unit of capacity smaller than the liter is the _____.

6. One metric unit of mass is the _____, and is equal to 1,000 grams.

7. A unit of mass smaller than the kilogram is the _____.

Concept Check

Circle the better estimate for each capacity.

8.

1 liter   1 milliliter

9.

1 liter   1 milliliter

How much water is in each container? Circle the amount.

10.    100 mL

200 mL

11.    400 mL

500 mL

Circle the better estimate for each mass.

12.    1 gram

1 kilogram

13.   1 gram

1 kilogram

Problem Solving

14. Mrs. Jones bought a 500 mL bottle of liquid laundry detergent. She used some of the detergent, and now she has 473 mL left. How much detergent did she use?

_____

15. **Algebra** Meg measured the mass of a banana and found it to be 132 grams. What is the total mass of 2 bananas of the same size? Write an equation with a letter for the unknown. Solve.

_____

Test Practice

16. Which is most likely to be the capacity of a mug of cocoa?

Ⓐ 10 milliliters   Ⓒ 2 liters

Ⓑ 100 milliliters   Ⓓ 5 liters

Tell Time to the Minute

ESSENTIAL QUESTION
Why do we measure?

Time can be represented on a digital clock or an analog clock.

A **digital clock** shows the time in numbers.

An **analog clock** has an hour hand and a minute hand.

Math in My World

Tools   Watch   Tutor

Example 1

The clock shows the time Erika got home from dance class. What time is it?

1 Find the hour.

The short hand is the hour hand. It has passed the 5.

But, it has not reached the 6. The hour is _____.

2 Count the minutes.

The longer hand is the minute hand.
Count by 5s first.

5, 10, 15, 20, 25, 30, 35, 40, _____
Then count each additional minute.

46, _____

There are _____ minutes past the hour.

Read: five forty-seven

Write: 5:47

So, Erika finished dance class at _____.

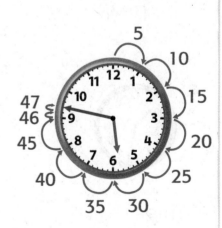

Example 2

Jake's soccer practice ended. What time is it?

The digits before the colon (:) show the hour.
The digits after the colon (:) show the minutes.

Read: twelve fifteen or
fifteen minutes past twelve

Write: _____

Guided Practice

Circle the correct time shown on each digital or analog clock.

1.

fifteen minutes before
five

five fifteen

2.

12:20

1:20

3.

2:45

1:50

4.

10:28

10:30

5.

three forty-four

forty-four minutes

Talk MATH

Does the minute hand or the hour hand move faster on an analog clock? Explain.

Independent Practice

Write the time shown on each digital or analog clock in numbers and words.

6.

Read: _____ minutes after _____

Write: _____

7.

Read: eleven _____

Write: _____

8.

Read: _____

Write: _____

9.

Read: _____

Write: _____

Draw a line to match each clock to its correct time.

10.

11.

4:02

7:17

3:45

8:53

12.

13.

Problem Solving

14. If the minute hand is pointing to the number 2, how many minutes past the hour is the clock showing?

_____

15. It was 12:53 when Jim's mom finished making lunch. Draw the hour hand and minute hand on the clock to show 12:53.

16. **Mathematical PRACTICE 3** **Draw a Conclusion** If the minute hand is pointing to the number 9, how many minutes before the next hour is the clock showing?

_____

HOT Problems

17. **Mathematical PRACTICE 5** **Use Math Tools** Write any time on the digital clock. Then describe in writing where the hour hand and minute hand would be pointing on an analog clock set at the same time.

_____

_____

18. **?** **Building on the Essential Question** Why is telling time important?

_____

_____

_____

MY Homework

Homework Helper   Need help? ⟋ connectED.mcgraw-hill.com

What time is shown on the clock?

1 **Find the hour.**
The hour hand is past the 3, but has not reached the 4.
The hour is 3.

2 **Count the minutes.**
Count by 5s first. Then count each additional minute.
5, 10, 15, 20, 25, 26
The minutes are 26.

Read: 3:26
Write: three twenty-six or twenty-six minutes after three

Practice

Circle the correct time shown on each digital or analog clock.

1.

6:35   7:35

2.

1:10   11:11

3.

7:03   7:05

4.

2:40   5:40

5.

4:50   5:50

6.

3:22   3:27

Write the time shown on each clock in numbers and words.

7.  Read: _____

   Write: _____

8.   Read: _____

   Write: _____

 Problem Solving

9. Diane's plane is due to arrive when the hour hand is just past the 3 and the minute hand is on the 4. What time should the plane arrive?

Mathematical
10. **PRACTICE** ➐ **Identify Structure** Clark was born when the hour hand was between 12 and 1 and the minute hand was on 11. What time was he born?

Vocabulary Check

Match each word or group of words to its meaning.

11. analog clock   • 60 seconds

12. digital clock   • a clock that shows the time with numerals and a colon

13. minute   • 60 minutes

14. hour   • a clock that shows the time with minute and hour hands

Test Practice

15. Look where the minute hand is pointing. How many minutes are after the hour?

   Ⓐ 25 minutes   Ⓒ 35 minutes

   Ⓑ 33 minutes   Ⓓ 38 minutes

Time flies!

Need more practice? Download Extra Practice at ⤳ **connectED.mcgraw-hill.com**

Time Intervals

To determine a **time interval,** find the amount of time that passes from the start of an activity to the end of an activity.

 Math in My World Tools Watch Tutor

Example 1

Evan's mother put biscuits in the oven at 5:10 P.M. She took them out of the oven at 5:27 P.M. How much time passed from the time she put the biscuits in the oven to the time she took them out?

Find the time interval from 5:10 to 5:27.

 Find the start time. The start time is _____ P.M.

 Find the end time. The end time is _____ P.M.

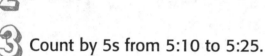 Count by 5s from 5:10 to 5:25.

 Count on by ones from 5:25 to 5:27.

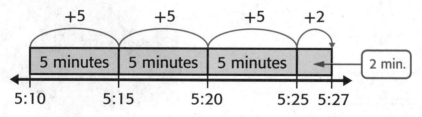

So, _____ minutes passed from 5:10 P.M. to 5:27 P.M.

The time interval was _____ minutes.

Example 2

The clock shows the time June's choir practice ended in the afternoon. If it was 135 minutes long, what time did it start?

Count back 135 minutes from 3:15 P.M.
Use the number line.

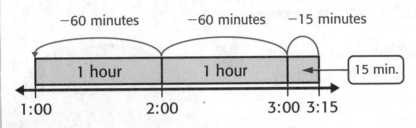

Helpful Hint
60 minutes = 1 hour

The time interval for choir practice was _____ minutes, or

_____ hours and _____ minutes.

135 minutes before 3:15 P.M. is _____ P.M.

Choir practice started at _____ P.M.

Guided Practice

1. Find the time interval of a movie. Use the number line.

Start Time (P.M.)   End Time (P.M.)

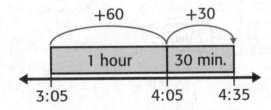

_____ minutes + _____ minutes = _____ minutes, or

_____ hour and _____ minutes.

Talk MATH

Lupe took a nap at the time shown and woke up at 1:30 P.M. Explain how you can find how long he slept.

666   Chapter 11 Measurement

Independent Practice

The following are times of baseball games. Find the time interval for each.

2. Start Time (P.M.) End Time (P.M.)

_____ minutes + _____ minutes + _____ minutes = _____ minutes

3. Start Time (P.M.) End Time (P.M.)

_____ minutes

4. Start Time (P.M.) End Time (P.M.)

_____ minutes

Write the time. Then draw the hands on the clock to show the time interval.

5. _____ A.M.

35 minutes earlier

_____ A.M.

6. _____ P.M.

36 minutes later

_____ P.M.

Problem Solving

7. Tyrone started mowing lawns at 1:10 P.M. He finished mowing at 4:05 P.M. How much time did Tyrone spend mowing lawns?

_____

8. It takes Louisa one hour and 30 minutes to get to her aunt's house. She leaves at 4:00 P.M. Draw the hands on the clock to show the time she will get to her aunt's house. Then write the time.

_____

HOT Problems

9. **Mathematical PRACTICE** 2 **Reason** How do the units of time, such as minutes and hours, relate to each other?

_____

10. **Mathematical PRACTICE** 3 **Which One Doesn't Belong?** Circle the set of digital clocks that does not show a time interval of 85 minutes. Explain.

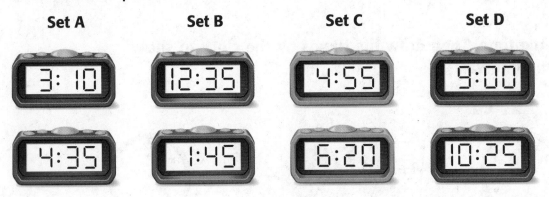

Set A	Set B	Set C	Set D
3:10	12:35	4:55	9:00
4:35	1:45	6:20	10:25

11. **Building on the Essential Question** How can I determine the duration of time intervals in hours?

_____

_____

MY Homework

Lesson 6
Time Intervals

Homework Helper

Need help? connectED.mcgraw-hill.com

Ramone took his pizza out of the oven at 2:50 P.M. The pizza had baked for 35 minutes. What time did Ramone put the pizza in the oven?

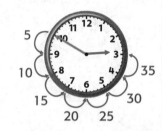

Start at 2:50 P.M. Counting by five-minute intervals, move backward around the clock until you count 35 minutes. The minute hand is on 3, and the hour is between 2 and 3.

So, Ramone put the pizza in the oven at 2:15 P.M.

Practice

The following are times of parades. Find the time interval for each.

1. Start Time (P.M.)   End Time (P.M.)

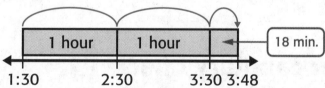

_____ minutes + _____ minutes + _____ minutes = _____ minutes

2. Start Time (P.M.)   End Time (P.M.)

_____ minutes

3. Start Time (A.M.)   End Time (A.M.)

_____ minutes

Write the time. Then draw the hands on the clock to show the time interval.

4. _____ P.M. 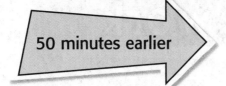 _____ P.M.

 50 minutes earlier

5. _____ A.M. 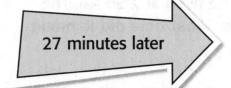 _____ A.M.

  27 minutes later

Problem Solving

6. It took Henry 2 hours and 17 minutes to write his report. He started writing at 3:30 P.M. What time did he finish his report?

_____

Mathematical
7. PRACTICE 1 **Plan Your Solution** Melanie started working in her garden at 8:25 A.M. She took a break at 11:10 A.M. How many minutes did Melanie work before taking a break?

_____

Vocabulary Check

8. Write a definition for a time interval. _____

_____

Test Practice

9. It takes Wallace 50 minutes to do the grocery shopping. If he starts at 2:12 P.M., what time will he finish?

Ⓐ 2:42 P.M. 　　Ⓒ 3:02 P.M.

Ⓑ 2:52 P.M. 　　Ⓓ 3:07 P.M.

Problem-Solving Investigation

STRATEGY: Work Backward

Learn the Strategy

Billy's picnic starts at 6:15 P.M. He needs to bake 3 dozen cookies. It takes 10 minutes to bake 1 dozen cookies. He will need another 20 minutes to get ready and 15 minutes to walk there. What is the latest time he can begin to prepare in order to be on time for the picnic?

1 Understand

What facts do you know?

the time the _____ starts

how long it takes to bake cookies, get ready, and _____ there

What do you need to find?
the latest time Billy can begin his preparations

2 Plan

Use the work backward strategy.

3 Solve

Use a number line. Mark everything in reverse order.

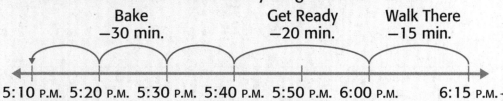

| Bake | Get Ready | Walk There |
| −30 min. | −20 min. | −15 min. |

5:10 P.M. 5:20 P.M. 5:30 P.M. 5:40 P.M. 5:50 P.M. 6:00 P.M.   6:15 P.M. ◄── Party Starts

Billy needs to start preparing by _____ P.M.

4 Check

Does your answer make sense? Explain.

Practice the Strategy

It took Liam 80 minutes to complete a jigsaw puzzle. He finished at 2:40 P.M. What time did he start?

 Understand

What facts do you know?

_____

_____

What do you need to find?

_____

2 Plan

_____

_____

3 Solve

_____

_____

_____

_____

_____

_____

4 Check

Does your answer make sense? Explain.

_____

Name _____

Apply the Strategy

Solve each problem by working backward.

1. Ava added an additional 35 milliliters of water to a vase. She then poured out 10 milliliters of water. Now she has 26 milliliters of water left in the vase. How much water was in the vase to begin with?

My Work!

2. Kenya left home for her aunt's house at 8:15 A.M. She arrived at her aunt's house 2 hours 50 minutes later. What time did she arrive? Write the time in words and numbers.

3. Ciro played at the park with 5 friends for 30 minutes, 7 friends for 1 hour, and 2 friends for 15 minutes. He arrived at the park at 1:00 P.M. What time did he leave the park?

4. Toya celebrated her birthday in March, 4 months after joining the swim team. Two months after joining the team, she swam in her first meet. What month did she swim in her first meet?

5. **Mathematical PRACTICE 1** **Keep Trying** Work through this puzzle to find the number of minutes that it took Logan to clean his room. Add 8 to a number. When you subtract 10 from the sum, and double the difference you get 44. What is the number?

Review the Strategies

Use any strategy to solve each problem.

- Draw a diagram.
- Determine reasonable answers.
- Work backward.
- Use models.

6. **PRACTICE** 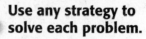 **Model Math** Blaine built a cube staircase. How many more cubes does he need to build 6 steps?

7. There are 3 children in line. Cami is right after Brock. Branden is third in line. What place is each child in line?

8. Flora, Alonso, and Luz went fishing. Find how many fish each caught.

Fish Caught

Name	Fish
Flora	3 more than Luz
Alonso	3 more than Flora
Luz	5 fish

9. Algebra A bottle of syrup has a mass of 560 grams. If the empty bottle has a mass of 305 grams, what is the mass of the syrup? Write an equation with a letter for the unknown. Then solve.

My Work!

MY Homework

Homework Helper eHelp

Need help? connectED.mcgraw-hill.com

Miranda will leave at 7:20 P.M. for a sleepover. She wants to be ready 5 minutes early. Miranda needs 45 minutes to do her chores, 20 minutes to shower, and 10 minutes to get dressed. What is the latest time Miranda can begin her chores?

1 Understand

What facts do you know?

the time Miranda will leave for a sleepover

how long it takes Miranda to complete each task

Miranda wants to be ready 5 minutes early.

What do you need to find?

the latest time Miranda can begin her chores

2 Plan

I will work backward to solve the problem.

3 Solve

Use a number line. Mark everything in reverse order.

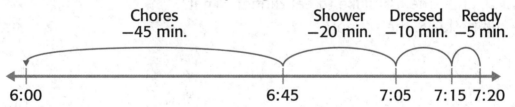

| Chores
−45 min. | | Shower
−20 min. | Dressed
−10 min. | Ready
−5 min. |

6:00   6:45   7:05   7:15 7:20

Miranda needs to start getting ready by 6:00 P.M.

4 Check

Does the answer make sense?

6:00 P.M. + 80 minutes = 7:20 P.M. So, the answer is correct.

Problem Solving

Solve each problem by working backward.

1. Selma finished her shift at work at 5:40 P.M. She had been working for 4 hours and 20 minutes. What time did Selma begin her shift?

2. Riko has 3 kilograms of potatoes left. She sold 8 kilograms at the farmers' market. She gave half that amount away to neighbors. How many kilograms of potatoes did Riko have to start with?

3. Noah earned $24 last week by pet sitting. He worked 1 hour on Friday and 3 hours on Wednesday. He worked the most hours on Monday. Noah earns $3 per hour. How many hours did he work on Monday?

4. **Mathematical** **PRACTICE** **6** **Be Precise** It took Blake 20 minutes to walk home from school. He spent 35 minutes doing homework. Then Blake played basketball for 1 hour and 10 minutes. Now it is 5:45 P.M. and time for Blake to eat dinner. What time did he leave the school?

5. Ms. Hirose filled a bucket with water. She used 3 liters of water to rinse her front porch. She used 2 liters of water to fill the bird bath. There are 3 liters of water left in the bucket. How much water did Ms. Hirose have to begin with?

My Work!

Vocabulary Check

Use the choices in the word bank to complete the puzzle.

analog clock	capacity	gram	kilogram
liter	mass	milliliter	time interval

ACROSS

3. a smaller metric unit of capacity than the liter

6. the amount a container can hold

7. one metric unit of capacity which is equal to 1,000 milliliters

8. one metric unit of mass which is equal to 1,000 grams

DOWN

1. the amount of time that passes from the start of an activity to the end of an activity

2. a metric unit for measuring mass, smaller than the kilogram

4. a clock that has an hour hand and a minute hand

5. the amount of matter, or material in an object

Concept Check

Circle the better unit to measure each capacity.

9.

milliliter

liter

10.

milliliter

liter

How much water is in each container? Circle the answer.

11.

1000 mL
800
600
400
200

550 mL

800 mL

12.

250 mL
200
150
100
50

100 mL

150 mL

Circle the better unit to measure each mass.

13.

gram

kilogram

14.

gram

kilogram

The following are times of after-school activities. Find the time interval for each.

15. Start Time (P.M.)   End Time (P.M.)

_____ minutes

16. Start Time (P.M.)   End Time (P.M.)

3:15   5:05

_____ minutes

Name

Problem Solving

My Work!

17. Mrs. Nathaniel bought 3 bottles of ice tea. Each bottle contains 500 milliliters of tea. How many milliliters of tea are there in 3 bottles altogether? Explain.

18. Algebra Ronaldo measured the mass of his soccer cleat to be 28 grams. The mass of his soccer ball is 450 grams. What is the mass of the soccer ball and soccer cleat together? Write an equation with a letter for the unknown. Then solve.

19. Circle the better estimate for the mass of this shark. Explain your choice.

80 grams   800 kilograms

Test Practice

20. Chia starts school at 8:30 A.M. It takes her 30 minutes to get dressed and 15 minutes to eat. Her walk to school is 5 minutes long. What is the latest time she can wake up and still be to school on time?

   Ⓐ 7:30 A.M.   Ⓒ 7:50 A.M.

   Ⓑ 7:40 A.M.   Ⓓ 8:00 A.M.

Reflect

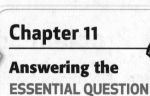

Use what you learned about measurement to complete the graphic organizer.

Vocabulary

Real-World Example: Mass

ESSENTIAL QUESTION

Why do we measure?

Real-World Example: Capacity

Real-World Example: Time

Now reflect on the ESSENTIAL QUESTION **Write your answer below.**

12 Represent and Interpret Data

How do we obtain useful information from a set of data?

MY Outdoor Adventures

Watch a video!

Watch ▶

MY Common Core State Standards

CCSS

Measurement and Data

3.MD.3 Draw a scaled picture graph and a scaled bar graph to represent a data set with several categories. Solve one- and two-step "how many more" and "how many less" problems using information presented in scaled bar graphs.

3.MD.4 Generate measurement data by measuring lengths using rulers marked with halves and fourths of an inch. Show the data by making a line plot, where the horizontal scale is marked off in appropriate units— whole numbers, halves, or quarters.

Operations and Algebraic Thinking *This chapter also addresses this standard:*

3.OA.3 Use multiplication and division within 100 to solve word problems in situations involving equal groups, arrays, and measurement quantities, e.g., by using drawings and equations with a symbol for the unknown number to represent the problem.

I'll be able to get this – no problem!

Standards for Mathematical PRACTICE

1. Make sense of problems and persevere in solving them.
2. Reason abstractly and quantitatively.
3. Construct viable arguments and critique the reasoning of others.
4. Model with mathematics.
5. Use appropriate tools strategically.
6. Attend to precision.
7. Look for and make use of structure.
8. Look for and express regularity in repeated reasoning.

= focused on in this chapter

Copyright © The McGraw-Hill Companies, Inc. Ingram Publishing/Image Source/SuperStock

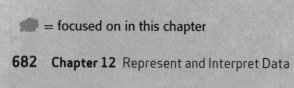

Name _____

Am I Ready?

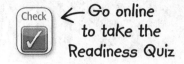 ← Go online to take the Readiness Quiz

Write the number represented by the tally marks.

1. |||

_____

2. ⊬||| |

_____

3. ⊬||| ⊬|||

_____

4. Mrs. Breeze surveyed her students to find their favorite pets. The results are shown. How many students voted for hamster?

_____

Favorite Pet	
Pet	**Tally**
Dog	⊬|||
Cat	|||
Hamster	⊬||| ||

5. The data shows the activities students enjoy in Physical Education class. How many more students voted for jump rope than kickball?

_____

Favorite Activity	
Jump rope	⌒ ⌒ ⌒ ⌒ ⌒
Basketball	● ● ●
Kickball	● ●
Volleyball	○

Identify a pattern. Then find the missing numbers.

6. 2, 4, 6, 8, ▦, ▦

_____

7. 5, 10, 15, 20, ▦, ▦

_____

8. 10, 20, 30, 40, ▦, ▦

_____

9. 100, 200, 300, ▦, ▦

_____

Shade the boxes to show the problems you answered correctly.

How Did I Do?

1	2	3	4	5	6	7	8	9

MY Math Words

Vocab
abc

Review Vocabulary

compare   symbol

Making Connections

Use one of the review vocabulary words to label the chart below.

Favorite Outdoor Adventures

Bird Watching	
Camping	
Fishing	

Use the information in the chart to compare camping and fishing.
Write a description. Then use >, <, or =.

MY Vocabulary Cards

Vocab

Lesson 12-2

analyze

hmmm.....

Lesson 12-3

bar graph

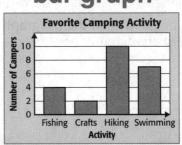

Favorite Camping Activity

Number of Campers / Activity: Fishing, Crafts, Hiking, Swimming

---

Lesson 12-1

data

Favorite Snack

Popcorn					
Carrots					
Apples					

Lesson 12-1

frequency table

How Do You Travel to School?	
Method	**Frequency**
Bicycle	3
Bus	6
Car	9
Walk	5

Lesson 12-6

half inch $\left(\dfrac{1}{2}\right)$

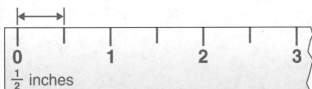

0   1   2   3

$\dfrac{1}{2}$ inches

Lesson 12-2

interpret

I learned that....

Lesson 12-2

key

Favorite Type of Movie

Comedy			
Drama			
Cartoon			

key: ☐ = 3 people

Lesson 12-5

line plot

Books Read in November

1   2   3   4   5

Ideas for Use

- During this school year, create a separate stack of cards for key math verbs, such as *analyze* and *interpret*. These verbs will help you in your problem solving.

- Draw or write examples for each card. Be sure your examples are different from what is shown on each card.

— —

A graph that compares data by using bars of different lengths or heights to show the values.

Explain why the bars in the bar graph shown on this card are different heights.

_____

_____

_____

To read and study the data on a graph.

Use the word *analyze* in a sentence. Be sure to use the word in a way that shows its meaning.

_____

_____

_____

A table that shows the number of times each result occurs.

The root word of *frequency* is *frequent.* Use it in a sentence.

_____

_____

Collected information or facts.

How could you collect the data used to make the graph shown on this card?

_____

_____

To explain what a graph shows.

Describe an example of when you need to interpret information at school.

_____

_____

One of two equal parts of one inch.

Describe when you might want estimate to the half inch.

_____

_____

_____

A graph that uses Xs above a number line to show how often a data value occurs.

How would you need to adjust the line plot on the front of this card to show that two students read 7 books?

_____

_____

Tells how many each symbol represents.

Why might a key be used to represent more than one item?

_____

_____

MY Vocabulary Cards

Lesson 12-2

pictograph

Favorite Pizza	
Cheese	●●
Pepperoni	●●●
Vegetable	●●(
key: ● = 2 students	

Lesson 12-2

picture graph

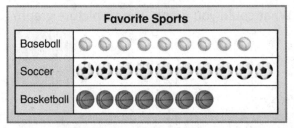

Lesson 12-6

quarter inch $\left(\dfrac{1}{4}\right)$

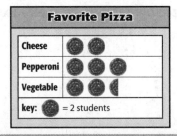

Lesson 12-3

scale

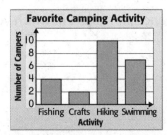

Lesson 12-1

survey

What Will You Do For Lunch Today?	
Lunch	Tally
Pack	卌 卌
Buy	卌 III
Buy Milk Only	IIII

Lesson 12-1

tally chart

Items Sold at School Store	
Item	Tally
Eraser	卌
Bottle of glue	卌 卌
Pencil	卌 III
Scissors	II

Lesson 12-1

tally mark(s)

卌 I

Ideas for Use

- Design a crossword puzzle. Use the definitions for the words as the clues.

- Use the blank cards to write a word from a previous chapter that you would like to review.

A graph that uses different pictures to represent each tally.

What could you graph using a picture graph?

_____

_____

A graph that uses the same symbol to represent more than one tally.

Does a pictograph need a scale? Explain.

_____

_____

A set of numbers that represents the data in a graph.

Describe another subject in which you use _scale._

_____

_____

One of four equal parts of one inch.

Quarter means "one-fourth." How does this relate to a quarter coin?

_____

_____

A chart that uses tally marks to show the results of data collection.

Look at the front of the card. How many total erasers, bottles of glue, pencils, and scissors were sold?

_____

To collect data by asking people the same question.

Write a question for a survey you might take of friends at a party.

_____

_____

A mark made to keep track and display data recorded from a survey.

Make tally marks to represent 14 votes.

_____

MY Foldable

FOLDABLES® Follow the steps on the back to make your Foldable.

Frequency Table

Bar Graph

What is your Favorite Outdoor Activity?

Pictograph

Tally Chart

FOLDABLES®
Study Organizer

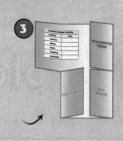

Favorite Outdoor Activity

Activity	Frequency				
Boating					
Hiking					
Picnic					
Sledding					
Swimming					

Favorite Outdoor Activity

Activity	Tally				
Boating					
Hiking					
Picnic					
Sledding					
Swimming					

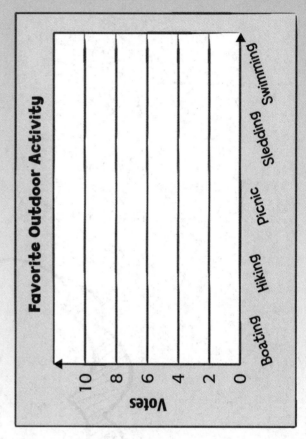

Favorite Outdoor Activity

Votes

10
8
6
4
2
0

Boating   Hiking   Picnic   Sledding   Swimming

Favorite Outdoor Activity

Boating

Hiking

Picnic

Sledding

Swimming

key =

Name _____

Collect and Record Data

Lesson 1
ESSENTIAL QUESTION
How do we obtain useful information from a set of data?

Data is collected information or facts. One way to collect data is by taking a **survey,** or asking a lot of people a question. A **frequency table** or **tally chart** will help you record the data you collect.

 ## Math in My World

Watch ▶ Tutor 💬

Example 1

Mr. Alvarez surveyed his scout troop. He asked each of his scouts, "What is your favorite camping activity?" The results are shown. Organize the data. Complete the table.

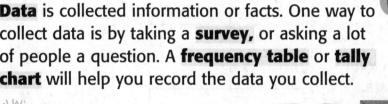

Swimming	Hiking	Fishing
Hunter	Amado	Julian
Eric	Avery	Chen
Ian	Omar	Lamarko
Jamal	Nicolas	
Alano		

One Way Use a tally chart.

Favorite Camping Activity	
Activity	**Tally**
Swimming	⍓⍓
Hiking	⍓⍓
Fishing	⍓

Show 5 with tallies.

Each **tally mark** represents one scout.

Another Way Use a frequency table.

Favorite Camping Activity	
Activity	**Frequency**
	5
Hiking	
Fishing	

Numbers are used to record the results.

Write one sentence about the data that was recorded.

_____

_____

Example 2

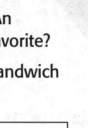
Open WIDE!

Survey your classmates. Record the results.

1. Write a survey question to ask your classmates. An example is shown. Which type of lunch is your favorite?

   A. grilled cheese   **C.** peanut butter and jelly sandwich

   B. pizza   **D.** spaghetti

2. Create a tally chart or a frequency table to record your results.

3. Ask the question to each of your classmates. Organize the data.

4. Write one sentence that describes your survey results.

Guided Practice

1. The data shows the sports cards most frequently traded in Mrs. Patton's class. Organize the data in a tally chart.

Traded Sports Cards	
Sport	**Tally**

Traded Sports Cards	
Sport	**Frequency**
Basketball	3
Baseball	6
Football	9
Hockey	5

Which is the most popular sports card to trade? _____

Which is the least popular? _____

Talk MATH

What is the difference between a frequency table and a tally chart?

Independent Practice

Organize each set of data in a frequency table.

2. While observing temperatures for one week, Arnaldo recorded the following data.

Weekly Temperatures	
Temperature (°F)	Tally
70–75	‖
76–80	‖|
81–85		
86–90		

Weekly Temperatures	
Temperature (°F)	Frequency

3. Darla observed her friends. She collected data on the flavor of milk they drank at lunch.

Flavors of Milk		
chocolate	strawberry	chocolate
white	white	strawberry
white	chocolate	strawberry
strawberry	white	chocolate
strawberry	strawberry	chocolate

Flavors of Milk	
Flavor	Frequency

Use the tally chart to answer the questions below.

4. Which item was the top seller?

5. Which item sold once?

6. How many items were sold altogether?

School Store Sales	
Item	Tally
Eraser	‖‖
Bottle of glue	
Pencil	‖‖ ‖|
Scissors		

Problem Solving

7. How would you represent the tallies at the right as a number in a frequency table?

$$\text{卌 卌 卌 II}$$

8. Elisa took a survey to find what breed of dog her classmates have. The results are shown. Record the data in the tally chart.

Dog Breeds	
beagle	beagle
golden retriever	beagle
golden retriever	golden retriever
poodle	beagle

Dog Breeds	
Dog	**Tally**

How many classmates responded to Elisa's survey? Explain.

How many more golden retrievers than poodles are owned

by Elisa's classmates? _____

HOT Problems

Mathematical
9. PRACTICE 5 Use Math Tools
Experiment with tossing a quarter, a nickel, dime, and penny 25 times each. Use the tally chart to record the number of times each coin lands heads up and tails up.

Experiment		
Coin	**Heads Up**	**Tails Up**

10. **Building on the Essential Question** What information can a tally chart and frequency table give you?

Measurement and Data
Preparation for 3.MD.3 and 3.MD.4

CCSS

MY Homework

Homework Helper

Need help? connectED.mcgraw-hill.com

While Ryan waited for the bus, he recorded the colors of the cars that went by in a tally chart. Write one sentence that describes the data.

Colors of Cars	
Color	**Tally**
Red	||||
Tan	ЖЖ ЖЖ
White	ЖЖ |
Blue	ЖЖ |

Ryan saw tan cars more than any other color car.

Practice

1. Alyssa records the results of her survey in the tally chart. Organize this data in the frequency table.

Favorite Day of School Week	
Day	**Tally**
Monday	||
Tuesday	||||
Wednesday	|
Thursday	ЖЖ |
Friday	ЖЖ ||||

Favorite Day of School Week	
Day	**Votes**

2. Which day of the school week is the least favorite day?

_____

Mathematical PRACTICE 4 **Model Math** For Exercises 3–5, refer to the following information. Fillipo observed the type of pants his friends wear to school.

3. Organize the set of data in a frequency table.

Pants Worn to School	
Type of Pants	Tally
Jeans	卌 I
Corduroys	IIII
Khaki	卌 II
Athletic	卌

Pants Worn to School	
Type of Pants	Frequency

4. How many more pairs of jeans were worn than corduroys? _____

5. Which type of pants was worn the most that day? _____

Vocabulary Check

Match each vocabulary term with its definition.

6. survey

7. frequency table

8. tally chart

- asking a group of people a question

- organizes the number of times each result has occurred in a table

- records the results of data collected using tally marks

Test Practice

9. According to the tally chart, how many students participated in the survey?

- Ⓐ 16 students
- Ⓑ 17 students
- Ⓒ 18 students
- Ⓓ 19 students

After School Activities	
Activity	Tally
Piano Practice	III
Soccer Practice	卌 IIII
Book Club	卌 II

Name ....

Draw Scaled Picture Graphs

Lesson 2
ESSENTIAL QUESTION
How do we obtain useful information from a set of data?

In Grade 2, you used a **picture graph** with different pictures to represent data. A **pictograph** uses the same symbol to represent more than one. For this reason, it is also called a scaled picture graph.

 Math in My World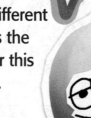

Example 1

Fifteen students drew pictures of their favorite fruits on sticky notes. Their responses are shown in the picture graph. How can this data be displayed in a pictograph?

Favorite Fruits

Banana						
Orange					
Strawberry			
Apple			

A pictograph displays the same data as a picture graph, but in a different way.

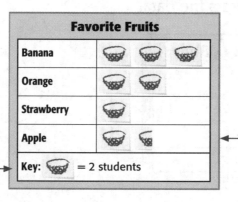

Favorite Fruits

Banana			
Orange		
Strawberry	
Apple		

Key: = 2 students

Make a key. The **key** tells how many each symbol represents.

A half basket represents _____ student.

There are half as many symbols in the _____ as the picture graph.

So, the graphs display the same set of data two different ways.

When you read a graph, you study, or **analyze,** the data. Then you are able to **interpret** the data, or explain what you learned.

Example 2

The pictograph shows the results of a survey Antoine conducted. Who saw two more movies than Grace?

Movies Seen During Summer Vacation

Zack	
Carla	
Grace	
Ivan	
Ricardo	
key: = 2 movies

The key shows that each symbol represents _____ movies.

Grace saw _____ + _____ + _____ or _____ movies.

2 + 2 + 2 = 6 movies

Carla saw _____ + _____ + _____ + _____ or _____ movies.

_____ + _____ + _____ + _____ = 8 movies

So, _____ saw 2 more movies than Grace.

Guided Practice

1. Analyze the pictograph. Then write a sentence that interprets the data.

Two Weeks of Weather

Sunny	☺ ☺ ☺
Cloudy	☺
Snow	☺ ☺
key: ☺ = 2 days

_____

_____

Talk MATH

Explain why a pictograph must have a key.

Name ....

Independent Practice

Display each set of data in a pictograph. Write a sentence that interprets the data.

2.

Sport Balls Sold on Saturday	
Type of Ball	**Frequency**
Football	6
Baseball	4
Basketball	7

Sports Balls Sold on Saturday

Football	
Baseball	
Basketball	

key: ◯ = 2 balls

3.

Fish Caught on Sunday	
Type of Fish	**Frequency**
Trout	10
Bass	8
Catfish	17

Fish Caught On Sunday

Trout	
Bass	
Catfish	

key: 🐟 = 2 fish

4. A barn had 6 of each animal shown below and 9 pigs.

Barn Animals

Cow	
Horse	
Pig	

key: **A** = 3 animals

Problem Solving

5. A pictograph shows 2 ♪ symbols. Each symbol represents 3 people who enjoy rock music. How many people enjoy rock music?

_____

6. Answer the questions about the pictograph. What is the most common shoe size?

_____

How many more students wear a size 4 shoe size than wear a size 8 shoe size?

_____

Third-Grade Shoe Sizes	
size 2	👟
size 4	👟 👟 👟
size 6	👟 👟 👟 👟 👟
size 8	👟
key: 👟 = 4 students	

Mathematical
PRACTICE 6 **Explain to a Friend** How many students were asked for their shoe size? Explain.

_____

_____

HOT Problems

Mathematical
7. PRACTICE 2 **Use Symbols** Ask 10 people which of the three states they would most like to visit. Record the data in a frequency table and display it in a pictograph.

States to Visit	
State	**Frequency**
New York	
California	
Florida	

States to Visit	
New York	
California	
Florida	
key: 👤 = 2 people	

8. **Building on the Essential Question** What is the difference between a picture graph and a pictograph?

_____

_____

MY Homework

Lesson 2

Draw Scaled Picture Graphs

Homework Helper

Need help? connectED.mcgraw-hill.com

**Eighteen people voted for their favorite instrument.
The results are shown in the two graphs below.
How many people voted for guitar?**

> A picture graph uses different pictures to represent each vote.

Favorite Instrument

Guitar	
Drums	
Trumpet	

> A pictograph, or scaled picture graph, uses the same symbol to represent more than one vote.

Favorite Instrument

Guitar	♩ ♩ ♩
Drums	♩ ♩ ♩ ♩
Trumpet	♩ ♩
key: ♩ = 2 votes	

> The key tells how many each symbol stands for.

♩ = 2, so 2 × 3 = 6

6 people voted for guitar.

Practice

1. Display the set of data in a pictograph. Then write a sentence that interprets the data.

How I Get to School	
Method	**Frequency**
Walk	5
Car	8
Bus	12

How I Get to School	
Walk	
Car	
Bus	
key: △ = 2 students	

2. Display the set of data in a pictograph. Write a sentence that interprets the data.

Dulaney Horse Farm	
Horse	**Frequency**
Appaloosa	16
Mustang	8
Clydesdale	10

Dulaney Horse Farm	
Appaloosa	
Mustang	
Clydesdale	
key: ∪ = 4 horses	

Problem Solving

3. PRACTICE Mathematical **2** **Understand Symbols** A pictograph key shows

3  symbols. Each symbol represents 7 hikers in the mountains. How many hikers are in the mountains altogether?

Vocabulary Check

Choose the correct word(s) to complete each sentence.

picture graph   pictograph   key   analyze   interpret

4. To _____, is to read or study the data on a graph.

5. The _____ in the graph tells how many each symbol stands for.

6. A _____ uses the same symbol to represent more than one vote or tally.

Test Practice

7. A pictograph key shows that each 🎥 symbol equals 6 movies. How many symbols equal 18 movies?

Ⓐ 2 symbols   Ⓒ 4 symbols

Ⓑ 3 symbols   Ⓓ 5 symbols

Need more practice? Download Extra Practice at ⟁**connectED.mcgraw-hill.com**

Draw Scaled Bar Graphs

Lesson 3

ESSENTIAL QUESTION
How do we obtain useful information from a set of data?

A **bar graph** uses bars of different lengths or heights to show data. In Grade 2, you used a bar graph with a scale of 1. A scaled bar graph can use a scale greater than 1.

April showers bring May flowers!

Math in My World

Example 1

Milo surveyed five grades to find the number of May birthdays. He recorded the data in a tally chart. Display the data in a vertical bar graph.

 Draw and label.
Draw a rectangle.

Label the side and bottom of the graph to describe the information.

Give the graph a title.

May Birthdays	
Grade	**Tally**
First	~~HHH~~
Second	~~HHH~~ ~~HHH~~ I
Third	III
Fourth	IIII
Fifth	II

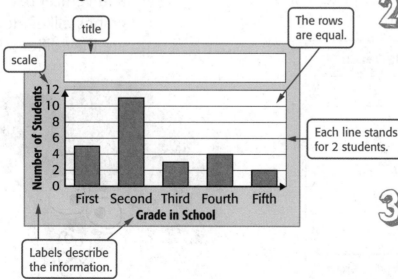

title

scale

The rows are equal.

Each line stands for 2 students.

Labels describe the information.

 Choose a scale.
A **scale** is a set of numbers that represents the data organized into equal intervals.

 Draw the bars.
Draw vertical bars to match the data.

Data can also be displayed in a horizontal bar graph, in which the bars go from left to right.

Example 2 Tutor

Desmond surveyed his friends about their favorite summer sports. Display the results in a horizontal bar graph.

1 Draw and label.
Draw a rectangle.

Label the side and bottom of the graph to describe the information.

Give the graph a title.

2 Choose a scale.
Write a scale on the bottom of the graph. Separate it into 6 equal columns.

3 Draw the bars.
Draw horizontal bars to match each number from your data.

Favorite Summer Sports		
Sport	Tally	Frequency
Tennis	IIII	
Swimming	HHT HHT	
Baseball	HHT II	
Biking	HHT I	

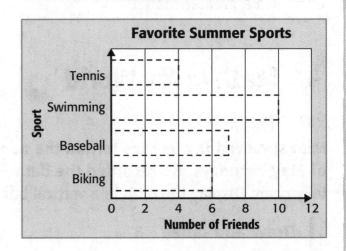

Guided Practice ✓ Check

1. Display the set of data from Example 2 in the vertical bar graph.

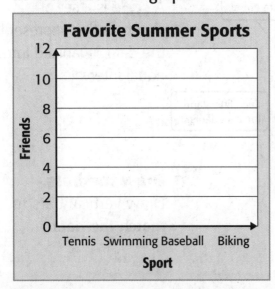

Talk MATH

How are horizontal and vertical bar graphs alike? How are they different?

Independent Practice

Display each set of data below in a bar graph.

2.

Favorite Birds to Watch	
Bird	**Tally**
Cardinal	卌 |||
Robin	||||
Goldfinch	卌 ||||

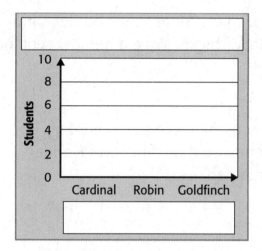

How many students responded to the survey altogether?

_____

How many more students like watching the goldfinch than the robin?

_____

3.

Animal Life Spans	
Animal	**Frequency**
Lion	10
Hamster	2
Kangaroo	5
Rabbit	7

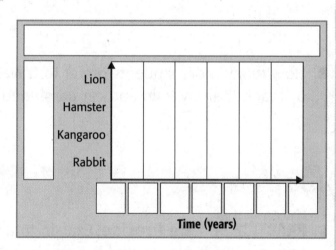

How many fewer years does the hamster live than the kangaroo?

_____

Which two animals when you subtract their life spans will equal the life span of the hamster?

_____

4. Write one sentence that describes the data in Exercise 3.

_____

Problem Solving

Mathematical PRACTICE 5 **Use Math Tools** Use the horizontal bar graph.

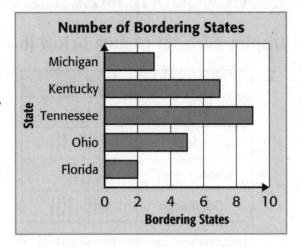

Number of Bordering States

5. How many more states border Ohio than Michigan? Write a number sentence.

_____

6. How many more states border Tennessee than Ohio and Florida combined?

_____

Use the vertical bar graph.

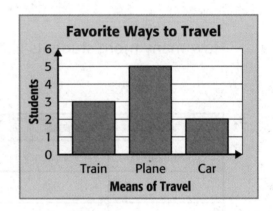

Favorite Ways to Travel

7. Write an addition equation to show the total number of students that were surveyed.

_____

8. How many more students prefer to travel by plane than by train and car combined?

_____

HOT Problems

Mathematical PRACTICE 6 **Be Precise** Explain to a friend the difference between the scales on the bar graphs on this page.

_____

_____

10. **Building on the Essential Question** What factors help you determine the scale you will use for a bar graph?

_____

_____

Name ....

MY Homework

Lesson 3

Draw Scaled Bar Graphs

Homework Helper

Need help? ↗ connectED.mcgraw-hill.com

Rodney surveyed his classmates to find which season they like best. First, he recorded the data in a tally chart. Then, he used the data to make a bar graph. How many more students voted for summer than for either spring or fall?

Favorite Seasons	
Season	**Tally**
Fall	‖
Winter	﷼HT l
Spring	‖‖
Summer	﷼HT ‖‖

Eight students voted for summer.
Four students voted for spring.
Two students voted for fall.

Find $8 - (4 + 2)$.

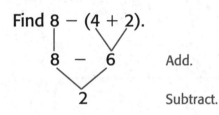

Add.

Subtract.

So, 2 more students voted for summer than for either spring or fall.

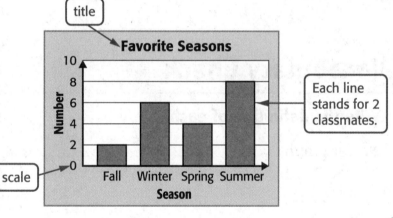

title

scale

Each line stands for 2 classmates.

Practice

1. Display the set of data below in a vertical bar graph.

Favorite Cities	
City	**Frequency**
New York	80
Denver	20
San Francisco	40
San Antonio	60

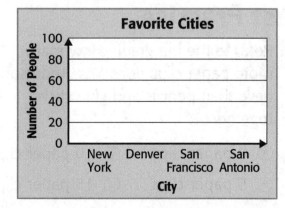

Mathematical PRACTICE 4 **Model Math** Display the data in a horizontal bar graph. Use the graph to answer Exercises 3–4.

2.

Language Spoken	
Language	**Tally**
French	|||
Spanish	ⵙⵙⵙ ⵙⵙⵙ
Chinese	ⵙⵙⵙ

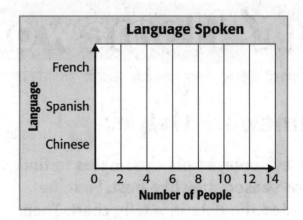

3. How many people participated in the survey?

_____

Mathematical
4. PRACTICE 2 **Use Number Sense** How many more people speak Spanish than Chinese and French combined?

_____

Vocabulary Check

Write the definition of each.

5. bar graph _____

_____

6. scale _____

_____

Test Practice

7. Refer to the bar graph. How many more paper clips does Mrs. Anderson have than pencils and glue sticks together?

Ⓐ 0 paper clips   Ⓒ 10 paper clips

Ⓑ 5 paper clips   Ⓓ 15 paper clips

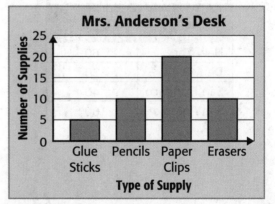

Relate Bar Graphs to Scaled Picture Graphs

Lesson 4

ESSENTIAL QUESTION
How do we obtain useful information from a set of data?

Math in My World

Example 1

Ishi counted the number of each type of fish in her fish tank. The scaled picture graph displays the data. Write a few sentences that interpret the data. Then answer the questions.

Ishi's Fish				
Jewelfish	🐟	🐟	🐟	🐟
Tetra	🐟	🐟		
Catfish	🐟			
Loach	🐟	🐟		
key: 🐟 = 2				

There are _____ jewelfish.

The number of catfish is _____ .

Ishi has _____ tetras and _____ loaches.

Altogether, Ishi has _____ fish.

What kind of fish does Ishi have the

fewest of? _____

Ishi has twice as many _____ as loaches.

Are there more or less jewelfish than catfish, tetras, and loaches combined? Explain.

_____

_____

Example 2 Tutor

Does the data change from when it is displayed in a scaled picture graph to when it is displayed in a bar graph?

Use the data from the scaled picture graph to complete the bar graph.

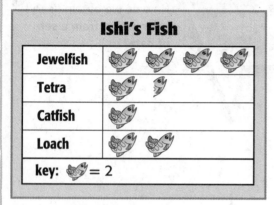

Ishi's Fish

Jewelfish	
Tetra	
Catfish	
Loach	

key: = 2

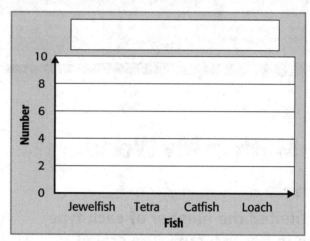

The scaled picture graph represents the data with _____.

The bar graph represents the data with _____.

Each scaled picture graph symbol stands for _____ fish. Each bar

graph scale interval stands for _____ fish.

So, the data does _____ change when it is displayed in a scaled bar graph.

Guided Practice Check ✓

1. Complete the bar graph using the data from the scaled picture graph.

Stars on Homework

Zoe	☆ ☆
Lyla	☆
John	☆ ☆ ☆ ☆

key: ☆ = 5

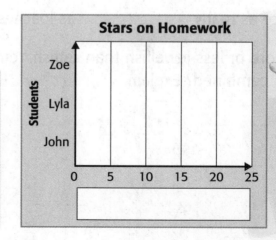

Talk MATH

If the scale for the bar graph above was in intervals of 4, would the information be different? Explain.

Independent Practice

Use the data from the first graph to display in the second graph. Write one sentence about the data in the graphs.

2.

Scaled Picture Graph

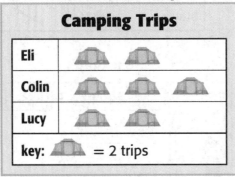

Bar Graph

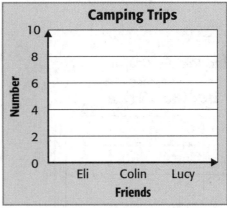

3.

Bar Graph

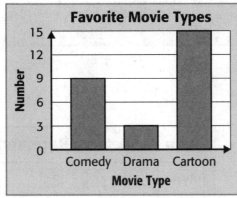

Scaled Picture Graph

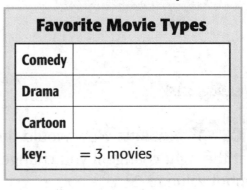

4. Complete the tally chart that may have been used to create this scaled picture graph.

Bodine's Tricks	
Trick	**Tally**
Sit	⊞⊞ ⊞⊞ IIII
Roll over	⊞⊞ III
Come	⊞⊞ ⊞⊞ II
Stay	⊞⊞ ⊞⊞ ⊞⊞ I

5. Ella kept a tally chart of all the times her dog did a trick. Display the data in a scaled picture graph and a bar graph.

Scaled Picture Graph

Bodine's Tricks	
Sit	
Roll over	
Come	
Stay	
key: 🐾 = 4 times	

Bar Graph

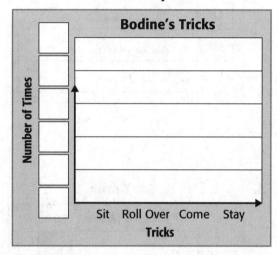

6. **Mathematical** **PRACTICE** 2 **Reason** How would the graph at the right change if you changed the symbol value to 2?

_____

_____

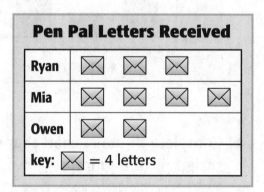

Pen Pal Letters Received

Ryan	✉	✉	✉	
Mia	✉	✉	✉	✉
Owen	✉	✉		
key: ✉ = 4 letters				

7. What might the scale for a bar graph that displays the same data with the symbol value of 4 be?

____ , ____ , ____ , ____ , ____

8. **Building on the Essential Question** Why are graphs helpful?

_____

_____

MY Homework

Homework Helper

Need help? connectED.mcgraw-hill.com

The bar graph shows the results of a teacher survey. Place the data from the bar graph into a scaled picture graph. Does the data change? Write one sentence about the data.

Bar Graph

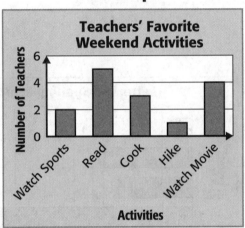

Scaled Picture Graph

The data is the same in both graphs. The graphs show reading is the preferred weekend activity of the teachers surveyed.

Practice

1. Complete the bar graph using the data from the scaled picture graph. Write one sentence about the data in the graphs.

Scaled Picture Graph

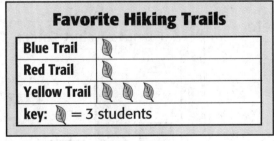

Bar Graph

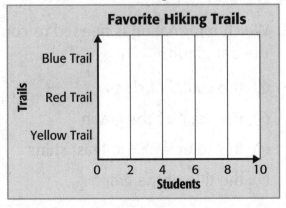

2. Complete the scaled picture graph using the data from the bar graph. Write one sentence about the data in the graphs.

Bar Graph

Scaled Picture Graph

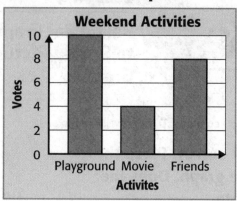

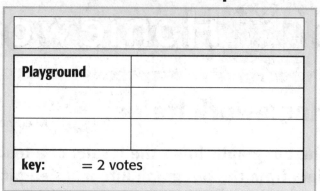

Problem Solving

The bar graph shows the number of letters in the third grade spelling words. Refer to the bar graph for Exercises 3–4.

3. How many letters do most of the spelling words have?

Mathematical
4. PRACTICE ➊ Make a Plan What might the key for a scaled picture graph that displays the same data be?

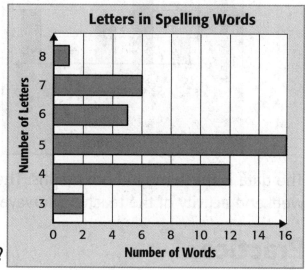

Test Practice

5. Which information is needed to complete the bar graph?

Ⓐ the colors of shirts

Ⓑ the scale of the graph

Ⓒ the color with the least shirts

Ⓓ the title of the graph

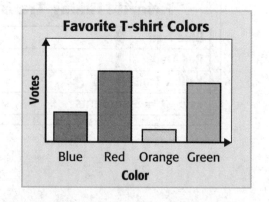

Draw and Analyze Line Plots

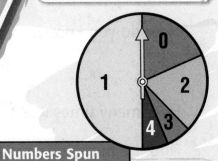

A **line plot** uses Xs above a number line to show how often a data value occurs.

  Math in My World Watch Tutor

Example 1

Albert spun a spinner 16 times to see how often the spinner landed on each number. Display the data in a line plot.

Numbers Spun			
0	1	2	1
1	2	0	1
2	1	4	1
0	1	1	3

1 Draw and label a line plot. Include all values of the data. Give it a title.

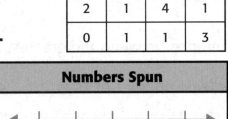

Include all values of the data. Use 0 to 4.

2 Draw an X above the number for each result. Complete the line plot by drawing the remaining Xs.

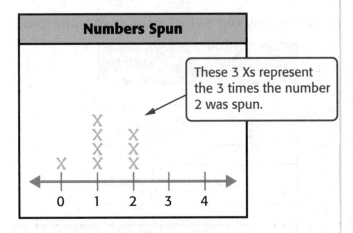

These 3 Xs represent the 3 times the number 2 was spun.

Analyze the line plot. Write a sentence that interprets the data.

_____

Example 2

Use Albert's line plot to find the difference between the greatest number of Xs and the least number of Xs.

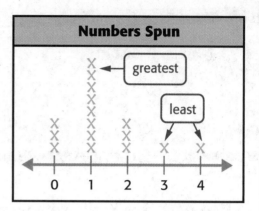

Numbers Spun

1 Find the number spun most often.

The number spun most often was _____ .

How do you know?

_____

How many times was the number 1 spun? _____

2 Find the numbers spun least often.

What numbers were spun least often? _____

How many times were each of these numbers spun? _____

3 The difference between the greatest number of Xs

and the least number of Xs is _____ − 1 = _____ .

Guided Practice

1. Display the set of data in the line plot.

Letters in Siblings' Names	
Number	Frequency
2	I
3	
4	II
5	II
6	I
7	II

Letters in Siblings' Names

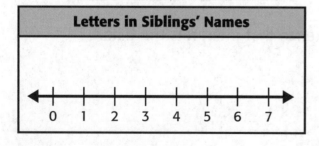

Talk MATH

Does a tally chart or a line plot make it easier to see how often numbers occur in a set of data? Explain.

Independent Practice

2. Display the set of data in the line plot.

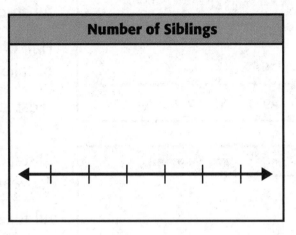

Number of Siblings	
Siblings	Frequency
0	1
1	3
2	5
3	3
4	2
5 or more	1

What is one conclusion you can draw from this line plot?

_____

_____

For Exercises 3 and 4, refer to the line plot below.

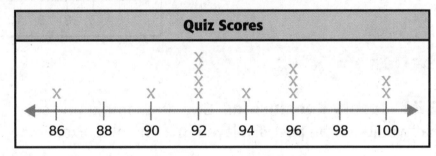

3. How many students' quiz scores are recorded? Explain.

_____

Mathematical
4. PRACTICE ③ Draw a Conclusion What is one conclusion
you can draw from this line plot?

_____

Problem Solving

Circle the true sentence about each set of data.

5.

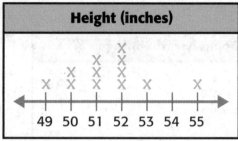

Height (inches)

49 50 51 52 53 54 55

All students are 55 inches tall.

Half of the students are 52 inches or taller.

Most students are 51 inches tall.

6.

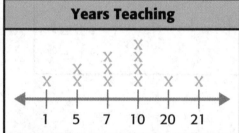

Years Teaching

1 5 7 10 20 21

All teachers have been teaching for 10 years or more, except 1.

All teachers have taught for 7 years.

Most teachers taught for 7 years or more.

7. Mathematical **PRACTICE** **2** **Use Symbols** Compare the number of members who are 6 years old and the number who are 8 years old. Use >, <, or =.

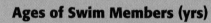
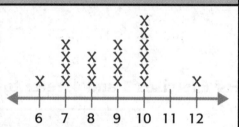

Ages of Swim Members (yrs)

6 7 8 9 10 11 12

_____

HOT Problems

8. Mathematical **PRACTICE** **3** **Justify Conclusions** Give an example of a set of data that would not be best displayed in a line plot. Explain.

_____

9. **Building on the Essential Question** How can I interpret the data I have collected?

_____

MY Homework

Homework Helper   Need help? connectED.mcgraw-hill.com

Jori recorded the number of birds she saw at the bird feeder each day for 5 days. She displayed the data in a line plot. What is one conclusion you can make from this line plot?

Number of Birds Seen Each Day	
Day	Tally
Day 1	|||
Day 2	||||
Day 3	||
Day 4	||
Day 5	|||

Number of Birds Seen Each Day

```
                X
        X       X
    X   X   X   X       X
    X   X   X   X       X
    X   X   X   X       X
  ◄─┼───┼───┼───┼───┼──►
    1   2   3   4   5
```

One conclusion that can be made is that there were 3 birds at the feeder, on 2 of the days.

Practice

Use the line plot above to answer Exercises 1–3.

1. How many days were there 2 birds at the bird feeder? _____

2. How many days were there 3 or fewer birds at the feeder? _____

3. How many days were there 3 or more birds at the feeder? _____

For Exercises 4 and 5, refer to the line plot that shows the number of states each student has visited.

4. How many states have the most number of students visited?

   _____

5. How many students have visited three states?

   _____

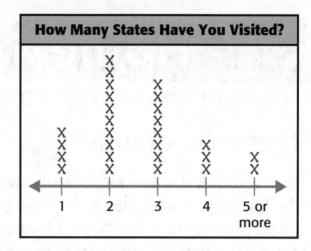

How Many States Have You Visited?

 Problem Solving

6. **Mathematical PRACTICE** **Model Math** Mrs. Sebring's class made a tally chart of the number of hours they spent on homework last week. Display the set of data in the line plot.

Weekly Time Spent on Homework	
Time (h)	Tally
8	III
9	I
10	HHI
11	III

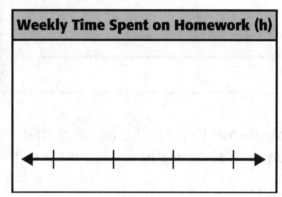

Weekly Time Spent on Homework (h)

Vocabulary Check

7. What is a line plot?

   _____

   _____

Test Practice

8. Refer to the line plot in Exercise 6. What is the difference between the least number of hours spent on homework and the most number of hours spent on homework?

   Ⓐ 1 hour   Ⓑ 3 hours   Ⓒ 8 hours   Ⓓ 11 hours

Check My Progress

Vocabulary Check

Use the word bank to label each graph, chart, or table.

bar graph   **frequency table**   **line plot**
pictograph   **picture graph**   **tally chart**

1.

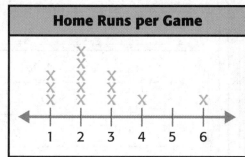

Home Runs per Game

2.

Favorite Type of Movie

Comedy	🖥 🖥 🖥
Drama	🖥
Cartoon	🖥 🖥 🖥 🖥 🖥

key: 🖥 = 3 people

3.

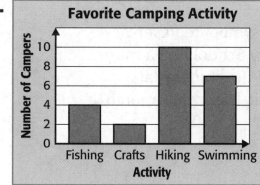

Favorite Camping Activity

4.

Spinner Results

Colors	Tally
Orange	‖‖‖ ‖‖‖
Red	‖‖‖
Green	‖‖

Use the remaining vocabulary words to complete each sentence.

5. A _____ uses numbers to record how often an event occurs.

6. Pictures are used as symbols in a _____ to show data.

Concept Check

7. Jack took a survey to find his friends' favorite type of book. Organize the data in a tally chart.

Favorite Type of Books	
Type of Book	Votes
Mystery	6
Science Fiction	7
Sports Stories	4

Favorite Type of Books	
Type of Book	Tally

8. Display the set of data from the picture graph in a scaled picture graph.

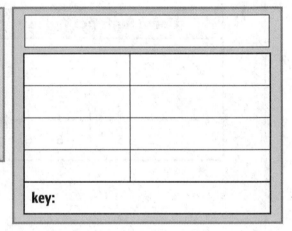

Stuffed Animal Prizes

Bear	
Cat	
Dog	
Turtle	

 # Problem Solving

9. A pictograph shows 8 ♪ symbols. Each symbol represents 5 times a student at Harrison Elementary School had a music lesson this month. How many times did the students at Harrison Elementary School have a music lesson this month?

Test Practice

10. According to the line plot, which statement is true?

Ⓐ Number 4 was rolled 2 times.

Ⓑ Number 6 was rolled the most.

Ⓒ The cube was rolled 20 times.

Ⓓ Number 1 was rolled the least.

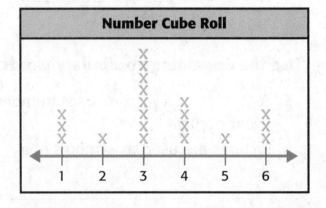

Measurement and Data

3.MD.4

CCSS

Hands On:
Measure to Halves and Fourths of an Inch

Lesson 6

ESSENTIAL QUESTION
How do we obtain useful information from a set of data?

One **half inch** $\left(\dfrac{1}{2}\right)$ is one of two equal parts of one inch. There are two half inches in one inch.

Measure It! Tools

Measure the length of three connecting cubes to the nearest half inch.

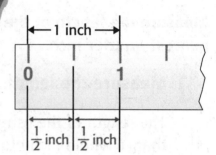

← 1 inch →

0   1

$\dfrac{1}{2}$ inch | $\dfrac{1}{2}$ inch

1 Measure the length.

Line up an inch ruler so that one end of the first cube is at the 0 mark.

The length of the three cubes is greater than 2 whole inches but

less than _____ whole inches.

It measures to the first mark after 2 inches.

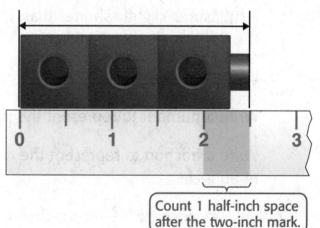

0   1   2   3

Count 1 half-inch space after the two-inch mark.

2 Write the length.

Write a number to represent the whole inches.

Write a fraction to represent the remaining part of an inch.

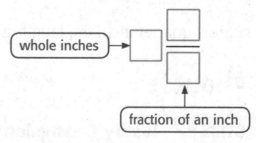

whole inches

fraction of an inch

The length is closer to $2\dfrac{1}{2}$ inches than it is to 2 inches or 3 inches.

So, to the nearest half inch, the length is ☐ $\dfrac{☐}{☐}$ inches.

Measuring to the nearest quarter inch allows you to be even more precise in your measurement.

One **quarter inch** $\left(\frac{1}{4}\right)$ is one of four equal parts of one inch.

Count the spaces, not the lines. There are 4 quarter inches in one inch.

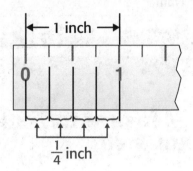

$\frac{1}{4}$ inch

Try It

Measure the length of the fish to the nearest quarter inch.

 Measure the length.

The length of the fish is greater than 1 whole inch but less than

_____ whole inches.

It measures slightly shorter than the third mark after 1 inch.

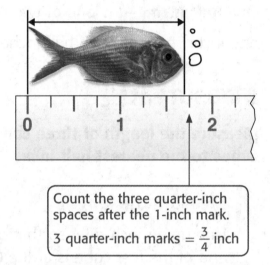

Count the three quarter-inch spaces after the 1-inch mark.

3 quarter-inch marks = $\frac{3}{4}$ inch

 Write the length.

Write a number to represent the whole inches.

Write a fraction to represent the remaining part of an inch.

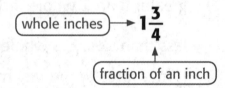
whole inches ⟶ $1\frac{3}{4}$
fraction of an inch

The length is closer to $1\frac{3}{4}$ inches than it is to $1\frac{1}{2}$ inches or 2 inches.

To the nearest quarter inch, the length is ☐ $\frac{☐}{☐}$ inches.

Talk About It

1. **Mathematical PRACTICE** ❸ **Justify Conclusions** Refer to the second activity. What is the length of the fish to the nearest half inch? How do you know?

_____

_____

Practice It

Estimate. Then measure each length to the nearest $\frac{1}{2}$-inch.

2.

Length: _____

3.

Length: _____

4.

Length: _____

Estimate. Then measure each length to the nearest $\frac{1}{4}$-inch.

5.

Length: _____

6.

Length: _____

7.

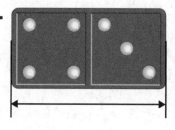

Length: _____

8.

Length: _____

Apply It

For Exercises 9 and 10, refer to the model airplane at the right. Fatima built the model to fly outdoors.

9. Measure the length of the airplane to the nearest half inch.

Length: _____

10. Measure the length of the airplane to the nearest quarter inch.

Length: _____

11. Mathematical **PRACTICE** 6 **Be Precise** Circle the object that, when measured to the nearest quarter inch, has a different length than when measured to the nearest half inch.

12. Mathematical **PRACTICE** 2 **Reason** Explain why the length of the paper clip to the nearest inch, half inch, and quarter inch are all the same number.

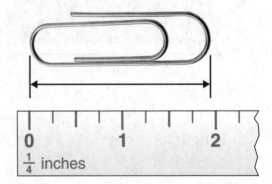

_____

_____

Write About It

13. Can an object be 3 inches in length when measured to both the nearest inch and nearest half inch? Explain.

_____

_____

MY Homework

Homework Helper [eHelp]

Need help? connectED.mcgraw-hill.com

Measure the length of the bolt to the nearest quarter inch.

1 Measure the length.

The length of the bolt is greater than 3 whole inches.

It measures even with the first mark after 3 inches.

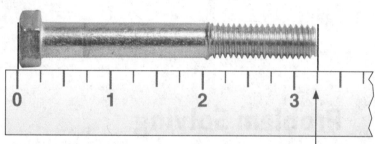

Count the one quarter-inch spaces after 3 inches.

one quarter-inch space $= \frac{1}{4}$ inch

2 Write the length.

Write a number to represent the whole inches.

Write a fraction to represent the remaining part of an inch.

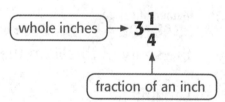

whole inches → $3\frac{1}{4}$

fraction of an inch

The length is closer to $3\frac{1}{4}$ inches than it is to 3 inches or $3\frac{1}{2}$ inches.

To the nearest quarter inch, the length is $3\frac{1}{4}$ inches.

Practice

1. Measure the length to the nearest $\frac{1}{2}$-inch.

Length: _____

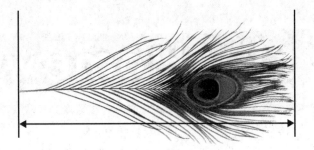

2. Measure the length to the nearest $\frac{1}{4}$-inch.

Length: _____

 ## Problem Solving

3. Measure the length of the marker to the nearest half inch.

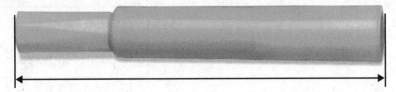

Length: _____

Mathematical
4. PRACTICE **6** **Be Precise** Measure the length of the spool of thread to the nearest quarter inch.

Length: _____

Vocabulary Check

5. How many half inches are in one inch? _____

6. How many quarter inches are in one inch? _____

Measurement and Data
3.MD.4, 3.OA.3

CCSS

Collect and Display Measurement Data

Lesson 7

ESSENTIAL QUESTION
How do we obtain useful information from a set of data?

You can gather measurement data by finding the lengths of several objects. The data collected can be useful.

 Math in My World [Tutor]

Example 1

Taylor is organizing a box of buttons by their size to the nearest half inch. Which size appears more often, $\frac{1}{2}$ inch or 1 inch?

Measure the buttons below to the nearest half inch. Organize the data in the tally chart, and then display it in a line plot.

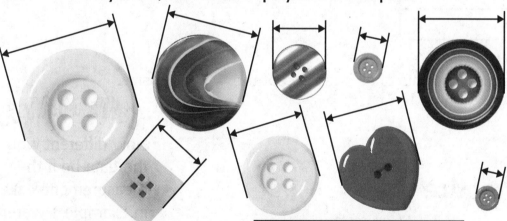

Buttons Sizes	
Length	**Tally**
$\frac{1}{2}$ in.	
1 in.	

Button Sizes

So, Taylor has more buttons that measure _____ inch than $\frac{1}{2}$ inch.

To be more precise, you can measure to the nearest $\frac{1}{4}$ inch.

Example 2

To the nearest $\frac{1}{4}$ inch, of which size are there more paper clips?

Measure each paper clip to the nearest $\frac{1}{4}$ inch. Organize the data in the tally chart, and then display it in a line plot.

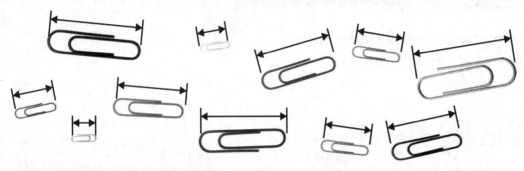

Paper Clips	
Length	Tally
$\frac{1}{4}$ in.	
$\frac{2}{4}$ or $\frac{1}{2}$ in.	
$\frac{3}{4}$ in.	
$\frac{4}{4}$ or 1 in.	

Paper Clips

0 $\frac{1}{4}$ $\frac{2}{4}$ or $\frac{1}{2}$ $\frac{3}{4}$ $\frac{4}{4}$ or 1

So, there are equal amounts of all sizes except the $\dfrac{\Box}{\Box}$ -inch size.

Talk MATH

How different would the data be if the measurements taken in Example 1 were to the nearest inch, instead of half inch?

Guided Practice

1. What do the Xs represent on the line plot in Example 2?

Independent Practice

Measure eight friends' pencils to the nearest inch.

2. Organize the data in a tally chart, and then display it in a line plot.

Pencil Lengths to the Nearest Inch	
Length	Tally

3. What is the difference in length between the longest and shortest pencil? Write an equation with the letter *p* for the unknown. Then solve.

4. What would be the total length of the pencils if they were laid end to end?

5. Use the same pencils. Measure them again but this time to the nearest half inch. Record and display your data.

Pencil Lengths to the Nearest Half Inch	
Length	Tally

6. How are the two sets of data different? How are they alike?

Problem Solving

7. Several students measured small items from home. The frequency table shows they measured the items to the nearest quarter inch. Display the data in the line plot.

Measurements	
Length	Frequency
$\frac{1}{4}$ in.	1
$\frac{2}{4}$ in.	4
$\frac{3}{4}$ in.	5
$\frac{4}{4}$ in.	2
$1\frac{1}{4}$ in.	2
$1\frac{2}{4}$ in.	3

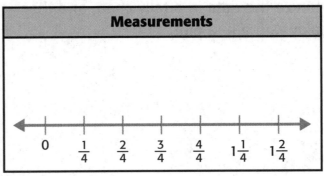

8. What is the total number of items that measured 1 inch or less?

9. **Mathematical PRACTICE** 2 **Use Algebra** Write an equation that shows the difference between the total items that measured 1 inch or less and the total items that measure more than 1 inch. Then find the difference.

HOT Problems

10. **Mathematical PRACTICE** 2 **Reason** How is a ruler like the number line on the line plot in Exercise 7?

11. **Building on the Essential Question** Why does recording measurement data lend itself to a line plot display?

Measurement and Data
3.MD.4, 3.OA.3

CCSS

MY Homework

Homework Helper

Need help? connectED.mcgraw-hill.com

Mrs. Carnes measured her students at the beginning of the school year and then again half way through the year. She recorded their growth in a tally chart. What is the most frequent measurement of growth of Mrs. Carnes' students?

Use a line plot to display the data.

Students' Growth in Third Grade	
Measurement	Tally
$\frac{1}{4}$ in.	|||
$\frac{2}{4}$ in.	||
$\frac{3}{4}$ in.	
$\frac{4}{4}$ in.	||||
$1\frac{1}{4}$ in.	||
$1\frac{2}{4}$ in.		

Students' Growth in Third Grade (inches)

```
              X
X             X
X   X         X
X   X         X   X
    |   |   | |   |   |
0   1   2   3   4   1 1  1 2
    4   4   4   4     4    4
```

The most frequent measurement of growth is $\frac{4}{4}$ of an inch, or 1 inch.

Practice

1. Tell why, in the line plot above, there are no Xs above $\frac{3}{4}$ of an inch.

   _____

   _____

2. Did more of Mrs. Carnes' students grow an inch or more or less than an inch? Explain.

   _____

   _____

3. Ribbon can be bought in many different widths. The tally chart shows how many spools of each width of ribbon can be bought. Represent the measurement data in a line plot.

Ribbon Widths	
Width	**Tally**
$\frac{1}{4}$ in.	|||
$\frac{2}{4}$ in.	||
$\frac{3}{4}$ in.	﹩
$\frac{4}{4}$ in.	||||
$1\frac{1}{4}$ in.		
$1\frac{2}{4}$ in.		

Ribbon Widths (inches)

```
 ◄———┼———┼———┼———┼———┼———┼———►
     0   1/4  2/4  3/4  4/4  1 1/4  1 2/4
```

Problem Solving

4. PRACTICE ② **Use Number Sense** Which two widths, of different length, when put together will equal $\frac{4}{4}$ of an inch or 1 whole inch?

5. Which two widths, of different length, when put together will equal $1\frac{1}{4}$ inches? Explain.

6. PRACTICE ① **Make Sense of Problems** Which has a greater combined width, two spools of ribbon that are each $\frac{2}{4}$ inch in width, or one spool that is $\frac{4}{4}$ inch in width? Explain.

Test Practice

7. How many spools of ribbon are represented in the line plot for Exercise 3?

ⓐ $1\frac{2}{4}$ spools   ⓒ 15 spools

ⓑ 14 spools   ⓓ 16 spools

Problem-Solving Investigation

STRATEGY: Solve a Simpler Problem

Lesson 8

ESSENTIAL QUESTION
How do we obtain useful information from a set of data?

Learn the Strategy

Shane rolls a 0–5 number cube and a 5–10 number cube together twenty times. The greatest possible sum is 15. Shane estimates that half of his rolls will have a sum of 15. Is his estimate reasonable?

1 Understand

What facts do you know?

the number of times he will roll the number cubes

Shane is estimating that half of the rolls will give a sum of _____ .

What do you need to find?

if his estimate is _____

2 Plan

I will collect and organize the data in a line plot. Then I will

decide if Shane's estimate is _____ .

3 Solve

Make a line plot. Roll the number cubes together. Record each sum with an X.

Half of 20 rolls are 10 rolls.

Shane only rolled 15 _____ times.

His estimate was _____ reasonable.

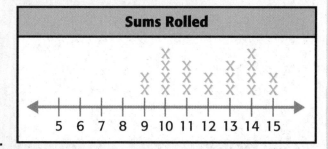

Sums Rolled

4 Check

Does your answer make sense? Explain.

_____

Practice the Strategy

Julina estimated that she needs to make 100 favors for the family reunion. Is this a reasonable estimate if 62 relatives come on Friday and half that many come on Saturday?

1 Understand

What facts do you know?

_____

_____

What do you need to find?

_____

_____

2 Plan

3 Solve

4 Check

Does your answer make sense? Explain.

_____

_____

Apply the Strategy

Solve each problem by first solving a simpler problem.

1. **Mathematical PRACTICE** ⑥ **Be Precise** Michael received a mixed-up box of T-shirts for the four teams he is coaching. He needs four of each number 1–5. He wrote down each number in a frequency table. Make a line plot to determine if he has enough of each number.

Team T-shirt Numbers			
1	3	1	5
4	5	4	2
2	2	1	3
3	1	5	3

Team T-shirt Numbers

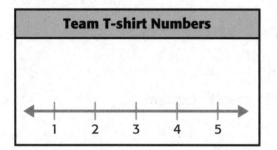

1   2   3   4   5

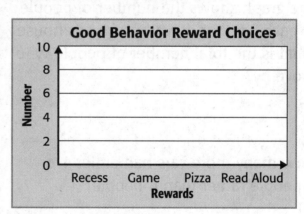

Does Michael have enough of each number? Explain.

_____

_____

2. Aubrey's class earned a reward for good behavior. The tally chart shows their votes. Put the data in a bar graph to determine if about half of the class voted for read-aloud time.

Good Behavior Reward Choices	
Reward	**Tally**
Extra recess	ⅢⅢ I
Game time	III
Pizza treat	ⅢⅢ III
Read-aloud time	ⅢⅢ ⅢⅢ

Good Behavior Reward Choices

Number: 10, 8, 6, 4, 2, 0

Recess   Game   Pizza   Read Aloud
Rewards

Is it reasonable to say that half the class voted for read-aloud time?

_____

_____

Review the Strategies

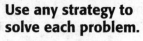

Use any strategy to solve each problem.

- Solve a simpler problem.
- Determine reasonable answers.
- Make a table.

3. Draw an example of a tally chart that may have been used to organize the data in the vertical bar graph below.

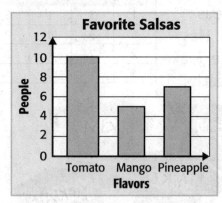

How many people were surveyed for this graph? Explain.

_____

_____

How many fewer people chose mango salsa than pineapple or tomato salsa combined?

_____

4. The graph shows the number of people in each car that drove by Niguel's house. What is the total number of people who drove by?

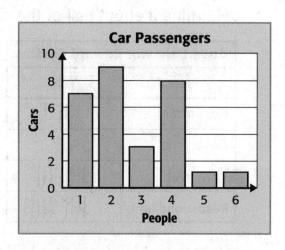

_____

How many more cars had either 1 or 2 people rather than 4 people?

_____

Mathematical PRACTICE 2 Reason Is it reasonable to say that about twice as many cars had 1 passenger than 3 passengers? Explain.

_____

738 Chapter 12 Represent and Interpret Data

Copyright © The McGraw-Hill Companies, Inc.

Name ....

MY Homework

Homework Helper

 eHelp

Need help? connectED.mcgraw-hill.com

Students recorded the number of letters in their first name in a tally chart. How many times as many students have 4 letters in their name than 3 letters?

First, solve a simpler problem by making a bar graph.

Number of Letters in our First Names	
Letters	**Tally**
3	$\mid\mid$
4	$\cancel{\mid\mid\mid\mid}\;\mid$
5	$\mid\mid\mid$
6	$\mid\mid\mid\mid$
7	$\mid$
10	$\mid$

Number of Letters in our First Names

Names (vertical axis): 0, 2, 4, 6, 8, 10
Letters (horizontal axis): 3, 4, 5, 6, 7, 10

Two names have 3 letters. Six names have 4 letters. So, 3 times as many students have 4 letters in their name than have 3 letters.

Problem Solving

Real World

1. What activity do Jaime's friends want to do most? Solve a simpler problem first by organizing the data in a tally chart.

Jaime's Party	
Activity	**Tally**

Alyssa – baseball
Jacob – swim
Kendra – picnic
Eve – swim
Luke – baseball
Kim – swim

Jaime and his friends will _____ .

Solve each problem by first solving a simpler problem.

2. Elizabeth surveyed her friends. She asked them to name their favorite sport. Is it reasonable to say that Elizabeth surveyed about 30 of her friends? First solve a simpler problem by making a bar graph.

Favorite Sports	
Sport	**Tally**
Soccer	~~HHT~~ I
Baseball	IIII
Football	~~HHT~~ III
Basketball	~~HHT~~ ~~HHT~~

3. **Mathematical PRACTICE ① Make Sense of Problems** Look at Exercise 2. What is the simpler problem you solved first?

4. How many more students chose football and baseball together than basketball? Write an equation.

5. **Mathematical PRACTICE ④ Model Math** Write a problem about the data above that would take two steps to solve. Then solve.

Vocabulary Check

Use the word bank below to complete the crossword puzzle.

analyze   bar graph   data   interpret   key

line plot   pictograph   scale   survey   tally chart

ACROSS

3. to explain what a graph shows

4. to collect data by asking people the same question

6. a graph that uses bars of different lengths or heights

7. to study data on a graph

DOWN

1. a scaled picture graph

2. tells "how many" each symbol represents

4. a set of numbers that represents the data

5. collected information

Concept Check ✓

For Exercises 8–13, refer to the table that shows the instrument each student plays.

8. Record the data in the tally chart.

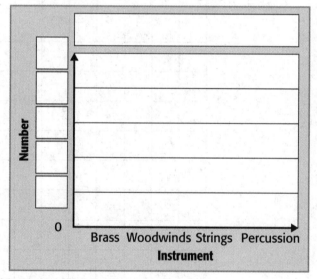

percussion	String	Brass	Woodwinds
Hunter	Amado	Julian	Lamarko
Eric	Avery	Chen	Aubrey
Ian	Omar	Luke	Isaac
Jamal	Matthew	Lily	Chase
Alano	Olivia		Isabelle
Jacob			
Hailey			

Orchestra Instruments	
Instrument	**Tally**

9. Display the set of data from above in the pictograph below.

Orchestra Instruments	

key: ♫ =

10. Display the set of data from the pictograph in the bar graph below.

Number

0

Brass   Woodwinds   Strings   Percussion

Instrument

11. How many students are not playing either a brass or a string instrument?

_____

12. Hunter is sick today. Miss Miller will practice with half of the students today. How many students will practice with Miss Miller today?

13. Which two instruments do an equal number of students play?

_____

Problem Solving

14. Matthew measured the length of some bugs. He displayed the data in the line plot. Find 5 of Matthew's errors.

Bug Length	
$\frac{1}{4}$ in.	1
$\frac{2}{4}$ in.	1
$\frac{3}{4}$ in.	5
$\frac{4}{4}$ in.	4
$1\frac{1}{4}$ in.	5
$1\frac{2}{4}$ in.	1

Bug Length

$$0 \quad \frac{1}{4} \quad \frac{2}{4} \quad \frac{3}{4} \quad \frac{4}{4} \quad \frac{1}{4} \quad \frac{2}{4}$$

Explain Matthew's errors.

_____

_____

15. Refer to Exercise 14. If the bugs that are $\frac{4}{4}$ inches in length were laid end to end, what would be their total length?

_____

16. There are three elementary schools on Brown Street: PS Elementary (PSE), PS Middle (PSM), and PS High (PSH).
• PSE is 1 mile from PSM.
• PSH and PSE are 2 miles apart.
• PSH is 3 miles from PSM.

Use the facts above. What is the order of the schools on the street?

_____

Test Practice

17. According to the line plot, how many pets do most students have?

   Ⓐ 1 pet   Ⓒ 3 pets

   Ⓑ 2 pets   Ⓓ 4 pets

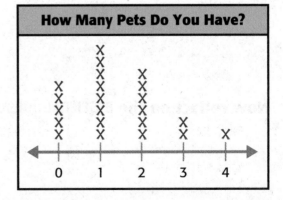

How Many Pets Do You Have?

Reflect

Use what you learned about data to complete the graphic organizer.

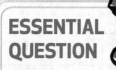

ESSENTIAL QUESTION

How do we obtain useful information from a set of data?

Real-World Problem

Bar Graph

Scaled Picture Graph

Now reflect on the ESSENTIAL QUESTION Write your answer below.

13 Perimeter and Area

ESSENTIAL QUESTION

How are perimeter and area related and how are they different?

Let's Build Something!

Watch a video!

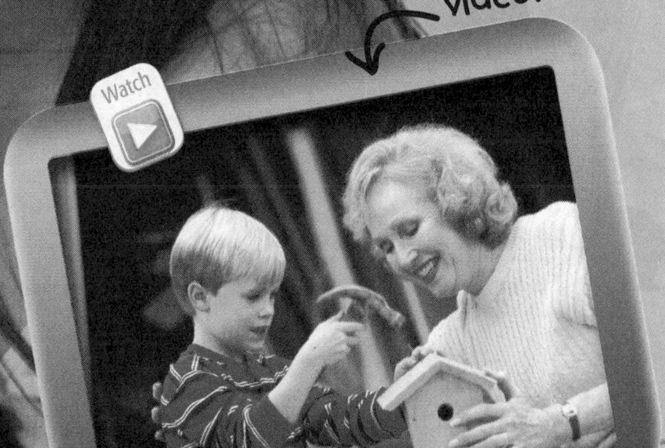

Watch ▶

MY Common Core State Standards

3.MD.5 Recognize area as an attribute of plane figures and understand concepts of area measurement.

3.MD.5a A square with side length 1 unit, called "a unit square," is said to have "one square unit" of area, and can be used to measure area.

3.MD.5b A plane figure which can be covered without gaps or overlaps by *n* unit squares is said to have an area of *n* square units.

3.MD.6 Measure areas by counting unit squares (square cm, square m, square in, square ft, and improvised units).

3.MD.7 Relate area to the operations of multiplication and addition.

3.MD.7a Find the area of a rectangle with whole-number side lengths by tiling it, and show that the area is the same as would be found by multiplying the side lengths.

3.MD.7b Multiply side lengths to find areas of rectangles with whole number side lengths in the context of solving real world and mathematical problems, and represent whole-number products as rectangular areas in mathematical reasoning.

3.MD.7c Use tiling to show in a concrete case that the area of a rectangle with whole-number side lengths a and $b + c$ is the sum of $a \times b$ and $a \times c$. Use area models to represent the distributive property in mathematical reasoning.

3.MD.7d Recognize area as additive. Find areas of rectilinear figures by decomposing them into non-overlapping rectangles and adding the areas of the non-overlapping parts, applying this technique to solve real world problems.

3.MD.8 Solve real world and mathematical problems involving perimeters of polygons, including finding the perimeter given the side lengths, finding an unknown side length, and exhibiting rectangles with the same perimeter and different areas or with the same area and different perimeters.

Standards for Mathematical PRACTICE

1. Make sense of problems and persevere in solving them.
2. Reason abstractly and quantitatively.
3. Construct viable arguments and critique the reasoning of others.
4. Model with mathematics.
5. Use appropriate tools strategically.
6. Attend to precision.
7. Look for and make use of structure.
8. Look for and express regularity in repeated reasoning.

= focused on in this chapter

Name

Am I Ready?

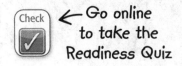 ← Go online to take the Readiness Quiz

Add.

1. 3 + 4 + 3 + 4 = _____

2. 5 + 6 + 5 + 6 = _____

3. 17 + 20 + 31 = _____

4. 40 + 63 + 12 = _____

5. From his house, Marcus walked three blocks north to the grocery store, six blocks east to the library, three blocks south to the park, and six blocks west back to his house. How many blocks did he walk in all?

_____

Multiply.

6. 3 × 5 = _____

7. 1 × 7 = _____

8. 4 × 6 = _____

9. 5 × 10 = _____

10. 8 × 9 = _____

11. 6 × 5 = _____

12. Write a multiplication sentence that represents the array shown at the right.

_____

Shade the boxes to show the problems you answered correctly.

How Did I Do?  | 1 | 2 | 3 | 4 | 5 | 6 | 7 | 8 | 9 | 10 | 11 | 12 |

MY Math Words

Review Vocabulary

decompose   Distributive Property

Making Connections

Use the review vocabulary to help you complete each section of the chart. Then answer the question.

What multiplication problem is represented by the shaded array?

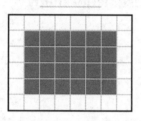

Show 8 × 5 in an array.

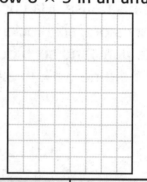

Decompose the above array.

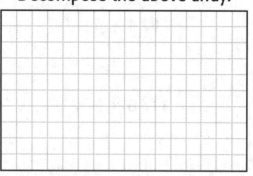

Decompose the 8 × 5 array.

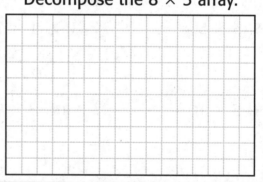

Explain how you used the Distributive Property to model 8 × 5.

....

....

MY Vocabulary Cards

Mathematical
PRACTICE

Lesson 13–3

area

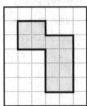

12 square units

Lesson 13–8

composite figure

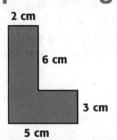

2 cm

6 cm

3 cm

5 cm

Lesson 13–6

formula

area (A) of a rectangle = length (ℓ) × width (w)

$$A = \ell \times w$$

Lesson 13–1

perimeter

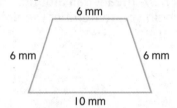

6 mm

6 mm   6 mm

10 mm

$P = 6\ mm + 6\ mm + 6\ mm + 10\ mm = 28\ mm$

Lesson 13–3

square unit

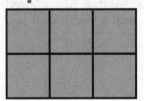

6 square units

Lesson 13–3

unit square

1 unit

Ideas for Use

- Group 2 or 3 common words. Add a word that is unrelated to the group. Then work with a friend to name the unrelated word.

- Use the blank cards to draw or write examples that will help you with concepts like relating area to multiplication and addition.

A figure made up of two or more figures.

Composite comes from the root word *compose,* meaning "to put together." Use *compose* to write a sentence that describes the figure on this card.

_____

_____

The number of square units needed to cover a figure without overlapping.

What is a real-life example of a situation when you would need to find area?

_____

_____

The distance around the outside of a figure.

The prefix *peri-* means "all around." How does this help you remember the definition of *perimeter?*

_____

_____

An equation that shows the relationship between two or more quantities.

Explain why formulas are helpful to use when finding area.

_____

_____

A square with a side length of one unit.

A friend tells you he measured his bedroom's perimeter in unit squares. Why is he incorrect?

_____

_____

A unit for measuring area.

Draw a rectangle that has a length of 6 square units and a width of 4 square units.

MY Foldable

FOLDABLES Follow the steps on the back to make your Foldable.

1 square unit	1 square unit	1 square unit	1 square unit	1 square unit	1 square unit
1 square unit	1 square unit	1 square unit	1 square unit	1 square unit	1 square unit
1 square unit	1 square unit	1 square unit	1 square unit	1 square unit	1 square unit
1 square unit	1 square unit	1 square unit	1 square unit	1 square unit	1 square unit
1 square unit	1 square unit	1 square unit	1 square unit	1 square unit	1 square unit
1 square unit	1 square unit	1 square unit	1 square unit	1 square unit	1 square unit

1

2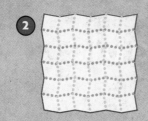

3

Measurement and Data

3.MD.8

CCSS

Hands On
Find Perimeter

Lesson 1

ESSENTIAL QUESTION
How are perimeter and area related and how are they different?

Perimeter is the distance around the outside of a figure, or shape. You can estimate and measure perimeter.

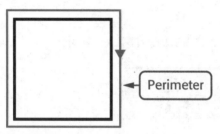

Perimeter

Measure It

1. Estimate the perimeter, in centimeters, of a piece of notebook paper. Record the results in the table below.

2. Use a centimeter ruler to find the perimeter to the nearest centimeter. Find the length of each side. Then add the side lengths. Record the results in the table.

3. Repeat Steps 1 and 2 for each object listed in the table.

Object	Perimeter	
	Estimate (cm)	Measure (cm)
piece of notebook paper		
math book		
desktop		
chalkboard or whiteboard		

An inch is larger than a centimeter. One inch is about halfway between 2 and 3 centimeters.

Try It

Using the same objects from the first activity, estimate and measure each perimeter to the nearest inch.

1 Estimate the perimeter, in inches, of a piece of notebook paper. Record the results in the table below.

2 Use an inch ruler to find the perimeter to the nearest inch. Record the results in the table.

3 Repeat Steps 1 and 2 for each object listed in the table.

Object	Perimeter	
	Estimate (in.)	Measure (in.)
piece of notebook paper		
math book		
desktop		
chalkboard or whiteboard		

Talk About It

1. **Mathematical PRACTICE** 6 **Explain to a Friend** Why is it important to estimate the perimeter before finding its exact measurement?

_____

2. Compare the perimeters of the objects in centimeters to the perimeters in inches. What do you notice?

_____

_____

3. After measuring the length of each side of an object, what operation did you use to find the perimeter? Explain.

_____

Name
....

Practice It

Estimate the perimeter of each figure in centimeters. Then measure the perimeter to the nearest centimeter.

4.

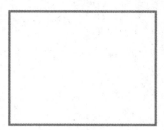

Estimate: _____

Actual: _____

5.

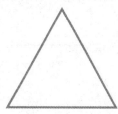

Estimate: _____

Actual: _____

Estimate the perimeter of each figure in inches. Then use an inch ruler to measure the perimeter to the nearest inch.

6.

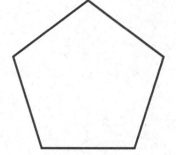

Estimate: _____

Actual: _____

7.

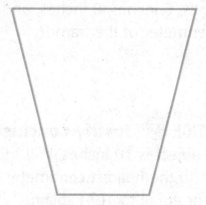

Estimate: _____

Actual: _____

8. Circle whether the number of inches used to measure the perimeter of a figure would be larger or smaller than the number of centimeters.

larger   smaller

 **Apply It**

Mathematical
9. PRACTICE 2 **Use Number Sense** Rebecca
used a centimeter ruler to measure the perimeter
of the figure below. Which estimate is closest
to the actual perimeter, 6 centimeters or
12 centimeters?

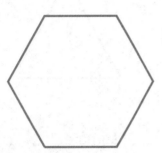

My Work!

10. Antoine built a picture frame out of wood. He used
an inch ruler to measure the perimeter of the
frame. If the lengths of the sides of the frame are
8 inches, 6 inches, 8 inches, and 6 inches, what is
the perimeter of the frame?

Mathematical
11. PRACTICE 3 **Justify Conclusions** The perimeter
of an object is 10 inches. Will the perimeter of the
object to the nearest centimeter be less than, greater
than, or equal to 10? Explain.

Write About It

12. How is perimeter related to the operation of addition?

MY Homework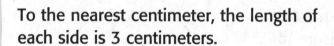

Homework Helper

Need help? connectED.mcgraw-hill.com

Use a centimeter ruler to measure the perimeter of the figure at the right to the nearest centimeter.

Measure the length of each side.

To the nearest centimeter, the length of each side is 3 centimeters.

Add the side lengths.

$3 + 3 + 3 + 3 = 12$

So, the perimeter of the figure is 12 centimeters.

Practice

Estimate the perimeter of each figure in centimeters. Then measure the perimeter to the nearest centimeter.

1.

Estimate: _____

Actual: _____

2.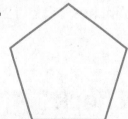

Estimate: _____

Actual: _____

Estimate the perimeter of each figure in inches. Then use an inch ruler to measure the perimeter to the nearest inch.

3.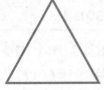

Estimate: _____

Actual: _____

4.

Estimate: _____

Actual: _____

 ## Problem Solving

Mathematical
5. PRACTICE ➡️ **Make Sense of Problems**
Gina used a centimeter ruler to measure the perimeter of the figure at the right. Which estimate is closest to the actual perimeter, 8 centimeters or 16 centimeters?

_____

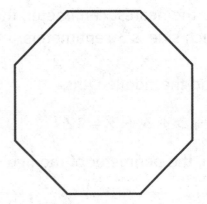

Mathematical
6. PRACTICE ➡️ **Be Precise** Allison used an inch ruler to measure the perimeter of the figure above. Circle the measure that represents the best estimate of the perimeter to the nearest inch.

2 inches   8 inches   12 inches   16 inches

Vocabulary Check

7. Complete the sentence below with the correct vocabulary word.

perimeter   array

_____ is the distance around a figure, or shape.

Perimeter

Lesson 2

ESSENTIAL QUESTION
How are perimeter and area related and how are they different?

The distance around the outside of a figure or shape is its **perimeter**.

 Math in My World Tools Watch

Example 1 Tutor

Warren and his dad will fence in the backyard for his new puppy. Find the perimeter of the backyard.

To find the perimeter if you already know the side lengths, add the lengths of the sides.

9 + 12 + 9 + 12 = _____

So, the perimeter is _____ meters.

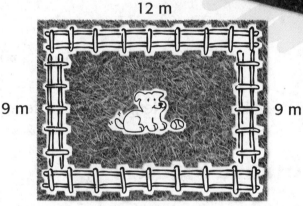

12 m

9 m   9 m

12 m

Example 2 Tutor

Find the perimeter of the shaded rectangle.

Count the distance around the figure, or add the lengths of the sides.

4 + 5 + 4 + 5 = _____

So, the perimeter is _____ units.

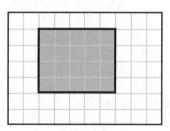

5 units

4 units   4 units

5 units

Key Concept Perimeter

Words	The perimeter of a figure is the distance around a figure, or the sum of its side lengths.

Model

4 cm

3 cm   3 cm

4 cm

Numbers

Perimeter = 3 cm + 4 cm + 3 cm + 4 cm
= 14 cm

Example 3

The perimeter of the figure is 33 feet. Find the unknown side length. Write an equation.

? ft

8 ft

9 ft

6 ft

6 ft

unknown

8 + 6 + 6 + 9 + ? = 33

29 + ? = 33   Add.

29 + 4 = 33   Think: 29 plus what number is 33?

The unknown side length is _____ feet, since 29 + _____ = 33.

Guided Practice

Find the perimeter of each figure.

1.

6 cm

2 cm   3 cm

5 cm

The perimeter is _____ centimeters.

2.
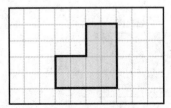

The perimeter is _____ units.

Talk MATH

If a triangle had three equal sides and its perimeter was 15 units, how could you find the length of each side?

Independent Practice

Find the perimeter of each figure.

3.

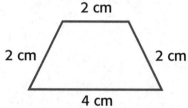

2 cm

2 cm   2 cm

4 cm

The perimeter is _____ centimeters.

4.

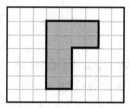

The perimeter is _____ units.

5.

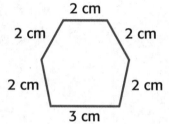

2 cm

2 cm   2 cm

2 cm   2 cm

3 cm

The perimeter is _____ centimeters.

6.

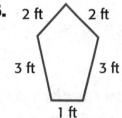

2 ft   2 ft

3 ft   3 ft

1 ft

The perimeter is _____ feet.

Algebra **Find the unknown side length for each figure.**
The perimeter of each figure is 50 centimeters.

7.

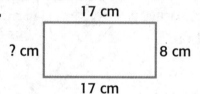

17 cm

? cm   8 cm

17 cm

The unknown is _____ centimeters.

8.

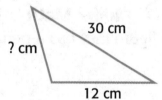

30 cm

? cm

12 cm

The unknown is _____ centimeters.

9.

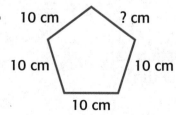

10 cm   ? cm

10 cm   10 cm

10 cm

The unknown is _____ centimeters.

10.

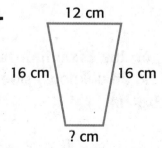

12 cm

16 cm   16 cm

? cm

The unknown is _____ centimeters.

Problem Solving

11. Maya's family is building a deck. The deck has 6 sides. Each side of the deck is 12 feet long. What is the perimeter of the deck?

12. The figure below has a perimeter of 21 feet. Find the length of the missing side.

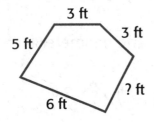

3 ft

3 ft

5 ft

? ft

6 ft

Mathematical
13. **PRACTICE** 2 **Use Algebra** A fountain has three sides. Its perimeter is 36 meters. One side is 12 meters and another is 15 meters. What is the length of the third side?

HOT Problems

Mathematical
14. **PRACTICE** 4 **Model Math** In the space below, draw and label a figure that has a perimeter of 24 inches.

15. **Building on the Essential Question** What operation can you use to find an unknown side length, if you know the perimeter? Explain.

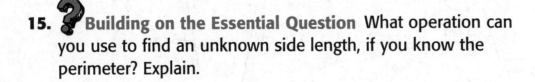

MY Homework

Homework Helper

Need help? connectED.mcgraw-hill.com

**The perimeter of the figure is 88 centimeters.
Find the unknown side length.**

10 cm
15 cm   ? cm
                  10 cm
15 cm
              10 cm
        15 cm

Write an equation.

unknown

$10 + 15 + 15 + 15 + 10 + 10 + ? = 88$

$75 + ? = 88$ Add.

$75 + 13 = 88$ Think: 75 plus what number is 88?

The unknown side length is 13 centimeters,
since $75 + 13 = 88$.

Check Add the lengths of all the sides.

$10 + 13 + 10 + 10 + 15 + 15 + 15 = 88$

Practice

Find the perimeter of each figure.

1.
10 cm
4 cm   4 cm
10 cm

2.
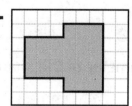

The perimeter is _____ centimeters. The perimeter is _____ units.

Algebra Find the unknown side length for each figure. The perimeter of each figure is 30 meters.

3.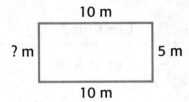

The unknown is _____ meters.

4.

The unknown is _____ meters.

 Problem Solving

Mathematical
5. **PRACTICE** 2 **Use Algebra** A garden has eight equal sides and has a perimeter of 56 meters. Circle the equation that gives the length, in meters, of each side.

$56 + 8 = 65$ $56 - 8 = 48$ $56 ÷ 8 = 7$

6. All professional baseball teams' playing fields are the same size. The three bases and home plate make a diamond that is 90 feet on each side. What is the perimeter of the diamond?

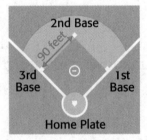

_____

Vocabulary Check

7. Describe perimeter in your own words.

_____

_____

Test Practice

8. What is the perimeter of the shaded figure?

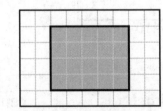

Ⓐ 18 units   Ⓒ 10 units

Ⓑ 20 units   Ⓓ 9 units

Name _____

Measurement and Data
3.MD.5, 3MD.5a, 3.MD.5b,
3.MD.6, 3.MD.8

CCSS

Hands On
Understand Area

Lesson 3

ESSENTIAL QUESTION
How are perimeter and area related and how are they different?

A square with a side length of one unit is called a **unit square**.

A unit square has one **square unit** of area and can be used to measure area. **Area** is the number of square units needed to cover a figure without overlapping.

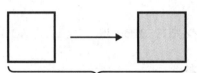

Shading or covering a unit square results in one square unit.

Draw It Tools

Draw and shade two different rectangles that each have an area of 20 square units.

Use the 10-by-10 grid.

To shade a rectangle with 20 square units, you need to shade a rectangle made up of 20 unit squares.

1. Shade 20 unit squares so that they form a rectangle.

   What is the perimeter of your rectangle?

   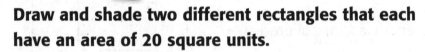

2. Shade another 20 unit squares so that they form a different rectangle.

   What is the perimeter of your rectangle?

   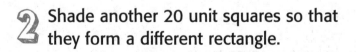

Online Content at ⟋ **connectED.mcgraw-hill.com**

You can also think of area as the amount of space enclosed by a figure.

Try It

Use a rubber band and a geoboard to make the rectangle shown. What is the area of the rectangle in square units?

How many unit squares are enclosed by the

rubber band? _____

So, the area is _____ square units.

Try It

What is the area of the figure at the right?

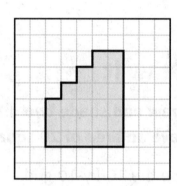

The figure has no gaps or overlaps. So, count the shaded unit squares.

How many unit squares are enclosed, or covered,

by the figure? _____

So, the area is _____ square units.

Talk About It

1. **Mathematical PRACTICE** ⑥ **Be Precise** Without drawing, tell how many different rectangles have an area of 5 square units. Explain.

_____

_____

_____

2. How can the term *unit square* help you to remember that area is measured in square units?

_____

Practice It

Count unit squares to find the area of each figure.

3.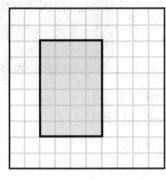

Area: _____

4.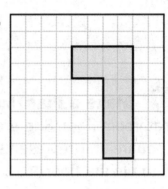

Area: _____

5. Draw and shade a rectangle with an area of 36 square units.

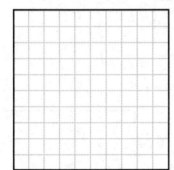

What is the perimeter of the figure you drew?

_____ units

6. Draw and shade a different rectangle with an area of 36 square units.

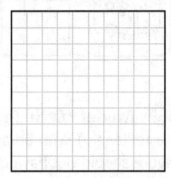

What is the perimeter of the figure you drew?

_____ units

7. A figure without gaps or overlaps can be covered by 14 unit squares. Circle the correct area of the figure.

4 square units   7 square units   14 square units

Apply It

8. Jared used a rubber band and geoboard to create the rectangle at the right. What is the area of the rectangle?

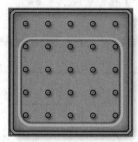

_____

Mathematical
9. PRACTICE 1 **Make a Plan** Morgan will help her parents tile a new bathroom floor. She drew a sketch of the bathroom floor. Each square unit represents one tile. How many tiles are needed to tile the floor?

Mathematical
10. PRACTICE 4 **Model Math** Draw and shade a figure (not a rectangle) with an area of 21 square units. The figure should not have any gaps or overlaps.

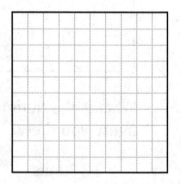

11. Find the perimeter of the figure you drew in Exercise 10.

Write About It

12. Describe one way that area can be measured.

768 Chapter 13 Perimeter and Area

Measurement and Data
3.MD.5, 3.MD.5a, 3.MD.5b, 3.MD.6, 3.MD.8

CCSS

MY Homework

Homework Helper

Need help? connectED.mcgraw-hill.com

What is the area of the figure at the right?

The figure has no gaps or overlaps. So, count the shaded unit squares.

There are 26 unit squares covering, or enclosing, the figure.

So, the area is 26 square units.

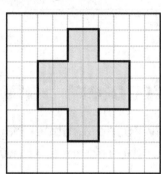

Practice

Count unit squares to find the area of each figure.

1.

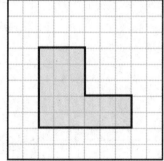

Area: _____

2.

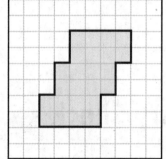

Area: _____

3. A shape is covered by 40 unit squares. What is the area of the shape?

_____

4. Draw and shade a rectangle with an area of 30 square units.

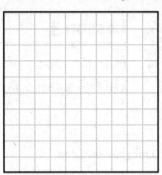

5. Draw and shade a different rectangle with an area of 30 square units.

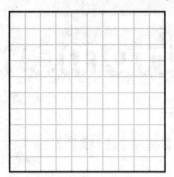

Problem Solving

6. Caitlyn used a rubber band and geoboard to make the rectangle shown. What is the area of the rectangle?

7. **Mathematical PRACTICE** **Plan Your Solution** A figure can be covered by 28 unit squares, without any gaps or overlaps. What is the area of the figure?

Vocabulary Check

Choose the correct word(s) to complete each sentence.

area   square units   unit square

8. _____ is measured in _____ and represents the number of those needed to cover a figure without overlapping.

9. A square with a side length of one unit is called a

_____ .

Measurement and Data

3.MD.5, 3.MD.5a, 3.MD.5b, 3.MD.6, 3.MD.7, 3.MD.8

CCSS

Measure Area

Lesson 4

ESSENTIAL QUESTION
How are perimeter and area related and how are they different?

Area is the number of square units needed to cover a figure without overlapping. Sometimes you need to count the number of half-square units covered by the figure.

Each of these is a $\frac{1}{2}$-square unit.

Math in My World

Tools Watch Tutor

Example 1

In art class, Hailey drew the figure at the right on grid paper. What is the area of the figure Hailey drew?

1 Count the number of whole squares.

There are _____ whole squares.

2 Count the number of half-squares.

There are 2 half-squares.
Two halves equal one whole.

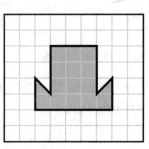

3 Add.

14 whole squares + 2 half-squares

14 whole squares + 1 whole square

_____ whole squares

So, the area is _____ square units.

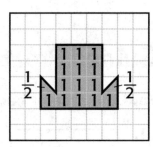

Sometimes, the units on drawings or figures represent another unit of measurement.

Example 2

Rafael created the geoboard figure at the right to represent a design he created. One square unit on the geoboard represents one square centimeter on the design. What is the area of the design?

1 Count the number of whole squares.

There are _____ whole squares.

2 Count the number of half-squares.

There are _____ half-squares. Eight halves equal four wholes.

3 Add.

8 whole squares + 8 half-squares

8 whole squares + 4 whole squares

_____ whole squares

So, the area is _____ square units.

The area of the design is _____ square centimeters.

Talk MATH

A figure is covered by 10 whole squares and some half-squares. If the area is 12 square units, how many half-squares are there? Explain.

Guided Practice

Find the area of each figure.

1.

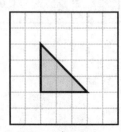

_____ square units

2.

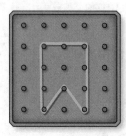

_____ square units

Independent Practice

Find the area of each figure.

3.

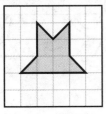

The area is _____ square units.

4.

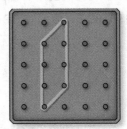

The area is _____ square units.

5.

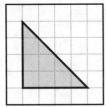

The area is _____ square units.

6.

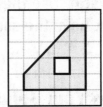

The area is _____ square units.

Find the area of each shaded region if one square unit represents one square inch.

7.

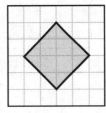

The area is _____ square inches.

8.

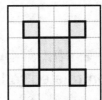

The area is _____ square inches.

9.

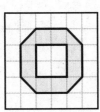

The area is _____ square inches.

10.

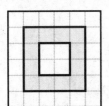

The area is _____ square inches.

Problem Solving

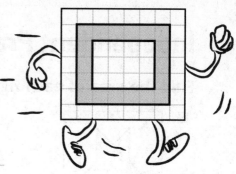

11. Denitra's family is building a stone walkway around their backyard. One square unit on the drawing at the right represents one square foot of the stone walkway. What is the area of the stone walkway?

_____

12. **Mathematical PRACTICE 5 Use Math Tools** Luisa is helping to tile a hallway. How many square tiles will be needed to fill the area?

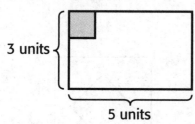

3 units

5 units

_____

HOT Problems

13. **Mathematical PRACTICE 2 Reason** Use the grid to draw two different figures that have the same area.

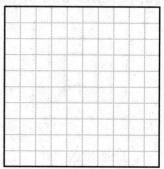

14. **Mathematical PRACTICE 1 Plan Your Solution** A rectangular room is 10 units wide by 14 units long. Find the area and perimeter of the room.

_____

15. **? Building on the Essential Question** How is the operation of addition related to finding area?

_____

Name ....

MY Homework

Lesson 4
Measure Area

Homework Helper

Need help? connectED.mcgraw-hill.com

Find the area of the figure at the right if each square unit represents one square centimeter.

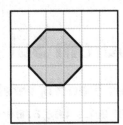

1 Count the number of whole squares.

There are 5 whole squares.

2 Count the number of half-squares.

There are 4 half-squares. Four halves equal two wholes.

3 Add.

5 whole squares + 4 half-squares

5 whole squares + 2 whole squares

7 whole squares

So, the area is 7 square units. If each square unit represents one square centimeter, then the area is 7 square centimeters.

Practice

Find the area of each figure.

1.

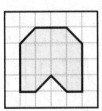

The area is _____ square units.

2.

The area is _____ square units.

Find the area of each shaded region if one square unit represents one square meter.

3.
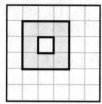

The area is _____ square meters.

4.
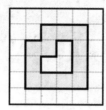

The area is _____ square meters.

Problem Solving

For Exercises 5 and 6, refer to the drawing at the right which represents the area of Elaine's bedroom.

5. What is the area of Elaine's bedroom in square units?

6. **Mathematical PRACTICE 8** **Look for a Pattern** If each square unit represents 5 square feet, what is the area of Elaine's bedroom in square feet? Use repeated addition.

Vocabulary Check

7. Describe area in your own words.

Test Practice

8. What is the area of the figure at the right?

Ⓐ 12 square units Ⓒ 14 square units

Ⓑ 13 square units Ⓓ 16 square units

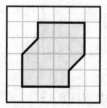

Check My Progress

Vocabulary Check

Fill in the correct word(s) that completes the sentence.

area   perimeter   square unit   unit square

1. The distance around a figure is its _____ .

2. A square with a side length of one unit is called a _____ .

3. _____ is measured in square units and represents the number of those needed to cover a figure without overlapping.

Concept Check

Estimate the perimeter of each figure in centimeters. Then measure the perimeter to the nearest centimeter.

4.

Estimate: _____

Actual: _____

5.

Estimate: _____

Actual: _____

Find the perimeter and area of each figure.

6.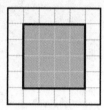

The perimeter is _____ units.

The area is _____ square units.

7.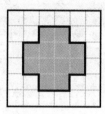

The perimeter is _____ units.

The area is _____ square units.

8. Algebra Find the unknown side length if the perimeter is 89 inches.

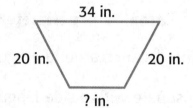

_____

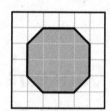

Problem Solving

Refer to the drawing at the right for Exercises 9 and 10.

9. Jeremy will help his father build a patio. The drawing represents the patio. What is the area of the patio in square units?

_____

10. If each square unit represents 3 square feet, what is the area of the patio in square feet? Use repeated addition.

_____

Test Practice

11. Each square unit on the figure represents one square meter. What is the area, in square meters, of the figure?

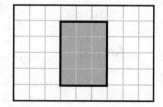

ⓐ 3 square meters   ⓒ 12 square meters

ⓑ 6 square meters   ⓓ 24 square meters

Measurement and Data

3.MD.5, 3.MD.5a, 3.MD.5b, 3.MD.6,
3.MD.7, 3.MD.7a, 3.MD.7b

CCSS

Hands On
Tile Rectangles to Find Area

Lesson 5

ESSENTIAL QUESTION
How are perimeter and area related and how are they different?

You can find the area of a rectangle on a grid by counting the number of unit squares. If a rectangle is not on a grid, you can find its area by tiling it.

The dimensions of a rectangle are its length and width.

Draw It

Find the area of the rectangle at the right by tiling it.

1 Tile the rectangle by separating the rectangle into unit squares. Draw unit squares on the rectangle so that the length of the rectangle is 8 unit squares and the width is 3 unit squares.

2 Count the total number of unit squares.

There are _____ unit squares.

So, the area of the rectangle is _____ square units.

Tiling the rectangle results in an array.

The array has _____ rows and _____ columns.

Find 3×8. $3 \times 8 =$ _____

What do you notice about the product of 3×8 and the total number of unit squares tiled in the rectangle?

_____

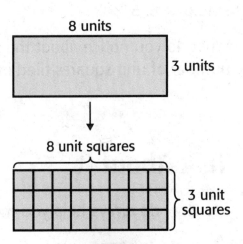

8 units

3 units

8 unit squares

3 unit squares

Try It

Find the area of the rectangle at the right by tiling it.

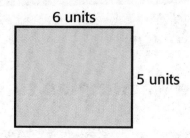

6 units

5 units

1 Tile the rectangle by separating the rectangle into unit squares. Draw unit squares on the rectangle so that the length of the rectangle is 6 unit squares and the width is 5 unit squares.

2 Count the total number of unit squares.

There are _____ unit squares.

So, the area of the rectangle is _____ square units.

Tiling the rectangle results in an array.

The array has _____ rows and _____ columns.

Find 5×6. $5 \times 6 =$ _____

What do you notice about the product of 5×6 and the total number of unit squares tiled in the rectangle?

Talk About It

1. How do arrays help you model the area of rectangles?

2. **Mathematical PRACTICE 2** **Stop and Reflect** How could you use the dimensions of a rectangle to find its area without tiling it?

3. Use your answer from Exercise 2 to find the area of a rectangle with a length of 7 units and a width of 4 units.

Practice It

Tile each rectangle to find its area. Draw unit squares on each rectangle.

4. 6 units

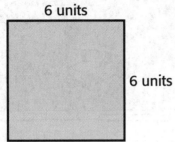

6 units

The area is _____ square units.

5. 5 units

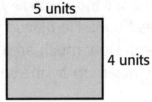

4 units

The area is _____ square units.

6. 4 units

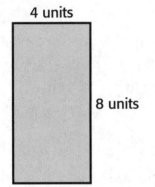

8 units

The area is _____ square units.

7. 3 units

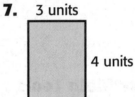

4 units

The area is _____ square units.

Algebra Find the area of each rectangle without tiling it. Write a multiplication equation.

8. 9 units

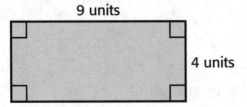

4 units

9. 7 units

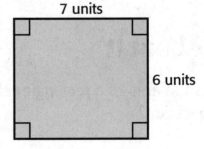

6 units

Apply It

Algebra **Write a multiplication equation to solve Exercises 10 and 11.**

10. Lucas built a sandbox for his younger brother. The length of the sandbox was 7 feet and the width was 5 feet. He placed the sandbox in the backyard. How much area of the backyard was taken up by the sandbox?

11. Perry created a one-page greeting card for his mom. The greeting card had a length of 6 inches and a width of 3 inches. What is the area of the greeting card?

12. **Mathematical PRACTICE 5** **Use Math Tools** In the space at the right, draw and tile a rectangle to represent Exercise 11.

13. **Mathematical PRACTICE 1** **Make Sense of Problems** A rectangle has a length of 8 meters and a width of 3 meters. Describe two different rectangles that have the same area as this rectangle.

Write About It

14. How is the area of a rectangle related to the operation of multiplication?

Measurement and Data

3.MD.5, 3.MD.5a, 3.MD.5b, 3.MD.6, 3.MD.7, 3.MD.7a, 3.MD.7b

CCSS

MY Homework

Lesson 5

Hands On: Tile Rectangles to Find Area

Homework Helper

Need help? connectED.mcgraw-hill.com

Find the area of the rectangle at the right by tiling it.

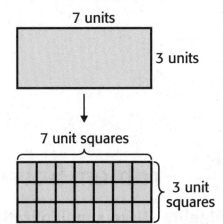

7 units

3 units

7 unit squares

3 unit squares

1. Tile the rectangle by separating the rectangle into unit squares. Draw unit squares so the length of the rectangle is 7 unit squares and the width is 3 unit squares.

2. Count the total number of unit squares.

   There are 21 unit squares.

So, the area of the rectangle is 21 square units.

Tiling the rectangle results in an array.
The array has 3 rows and 7 columns.
Find 3×7. $3 \times 7 = 21$

The product of 3×7 and the total number of unit squares tiled in the rectangle are the same.

Practice

Tile each rectangle to find its area. Draw unit squares on each rectangle.

1.
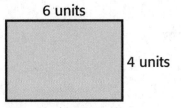

6 units

4 units

The area is _____ square units.

2.
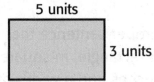

5 units

3 units

The area is _____ square units.

Algebra Find the area of each rectangle without tiling it. Write a multiplication equation.

3.

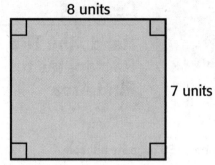

8 units

7 units

_____

4.

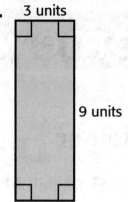

3 units

9 units

_____

 Problem Solving

Algebra Write a multiplication equation to solve Exercises 5 and 6.

5. A piece of poster board is in the shape of a rectangle. The length of the poster board is 2 feet and the width is one foot. What is the area of the piece of poster board?

_____

6. **Mathematical** **PRACTICE** **4** **Model Math** A rectangular garden has a length of 8 meters and a width of 5 meters. What is the area of the garden?

_____

7. Circle the number sentence that correctly represents the area of a rectangle, in square inches, with a length of 4 inches and a width of 10 inches.

$4 + 10 = 14$ $\qquad$ $4 \times 10 = 40$ $\qquad$ $4 + 10 + 4 + 10 = 28$

My Work!

Name _____

Measurement and Data

3.MD.5, 3.MD.5a, 3.MD.5b, 3.MD.6, 3.MD.7, 3.MD.7a, 3.MD.7b, 3.MD.8

CCSS

Area of Rectangles

Lesson 6

ESSENTIAL QUESTION
How are perimeter and area related and how are they different?

 Math in My World

Tools Watch Tutor

Example 1

A park manager is building a small rectangular playground. It will be 10 meters by 7 meters. Its area will be covered with shredded tires. What is the area of the playground that will be covered with shredded tires?

One Way Tile a rectangle.

1. Tile a rectangle with unit squares. It is 10 unit squares long and 7 unit squares wide.

   Each unit square represents one square meter.

2. Count the unit squares.

   There are _____ unit squares.

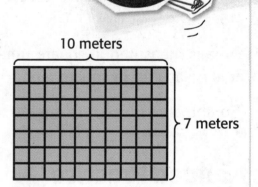

10 meters

7 meters

Another Way Multiply the side lengths.

Multiply the length by the width.

$10 \times 7 =$ _____   The length is 10 meters and the width is 7 meters.

Area is measured in square units. In this case, it is measured in square meters.

So, the area of the playground is _____ square meters.

Check You can check by using repeated addition to count the number of squares in each row.

$$10 + 10 + 10 + 10 + 10 + 10 + 10 = \text{_____}$$

A **formula** is an equation that shows the relationship between two or more quantities. A formula uses letters to represent the quantities. You can use a formula to find the area of a rectangle.

Key Concept Area of a Rectangle

Words To find the area A of a rectangle, multiply the length ℓ by the width w.

Formula $A = \ell \times w$

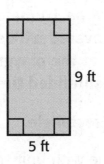

Example 2

Find the area of the rectangle.

Replace each symbol with its value.

$A = \ell \times w$ Use the area formula.
$A = 9 \times 5$ The length is 9 feet and the width is 5 feet.
$45 = 9 \times 5$ Multiply.

Area is measured in square units. In this case, it is measured in square feet.

So, the area is _____ square feet.

9 ft
5 ft

Guided Practice

Find the area of each rectangle.

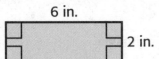

Talk MATH

Explain two ways to find the area of a rectangle.

1.   6 in.
2 in.

The area is _____ square inches.

2.   3 m
3 m

The area is _____ square meters.

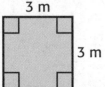

Independent Practice

Find the area of each rectangle.

3. 4 ft

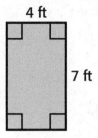
7 ft

_____ square feet

4. 5 m

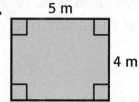
4 m

_____ square meters

5. 8 in.

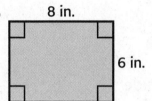
6 in.

6. 10 cm

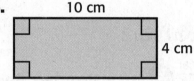
4 cm

7. A rectangle has an area of 42 square meters. Which could represent the length and width of this rectangle? Circle it.

7 meters and 6 meters   6 meters and 8 meters

Algebra Find the unknown side. Use the area formula.

8. ℓ

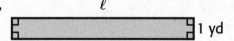
1 yd

The area is 9 square yards.

$A = \ell \times w$

$9 = $ _____ $\times 1$

The unknown is _____ yards.

9. 3 ft

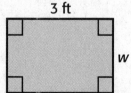
w

The area is 6 square feet.

$A = \ell \times w$

$6 = 3 \times$ _____

The unknown is _____ feet.

10. Mathematical **PRACTICE** 5 **Use Math Tools** Andrea is making a quilt in the shape of a rectangle. The length of the quilt will be 6 feet and the width will be 5 feet. What is the area of the quilt she will make? Write an equation to solve.

_____

11. A throw rug is 8 feet long and 4 feet wide. It is in a rectangular room with an area of 110 square feet. How much of the room is *not* covered by the rug?

_____

HOT Problems

12. Mathematical **PRACTICE** 2 **Use Number Sense** A rectangle has side lengths of 5 centimeters and 3 centimeters. If the side lengths are doubled, will the area also double? Explain.

_____

_____

13. Mathematical **PRACTICE** 4 **Model Math** Draw and label two rectangles that each have an area of 24 square inches, but have different perimeters.

14. **? Building on the Essential Question** How can multiplication and division be used to solve problems involving the area of rectangles?

_____

_____

Name

MY Homework

Lesson 6

Area of Rectangles

Homework Helper

Need help? connectED.mcgraw-hill.com

Find the area of a rectangle with a length of 8 inches and a width of 7 inches.

One Way Tile a rectangle.

1. Tile a rectangle with unit squares. It is 8 unit squares long and 7 unit squares wide.

   Each unit square represents one square inch.

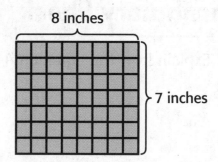

8 inches

7 inches

2. Count the unit squares.
   There are 56 unit squares.

Another Way Use $A = \ell \times w$.

$A = \ell \times w$ Area formula

$A = 8 \times 7$ The length is 8 inches and the width is 7 inches.

$56 = 8 \times 7$ Multiply.

Area is measured in square units. In this case, it is measured in square inches. So, the area is 56 square inches.

Practice

Find the area of each rectangle.

1. 4 ft

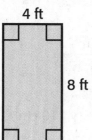

8 ft

_____ square feet

2. 7 m

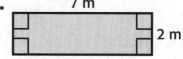

2 m

_____ square meters

Algebra Find the unknown side. Use the area formula.

3.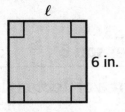
ℓ
6 in.

$36 = ℓ \times w$

$36 = \underline{\hspace{1.5cm}} \times 6$

The unknown is _____ inches.

4.
8 m
w

$24 = ℓ \times w$

$24 = 8 \times \underline{\hspace{1.5cm}}$

The unknown is _____ meters.

Vocabulary Check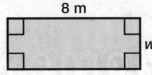

5. Explain how the equation $A = ℓ \times w$ is a formula.

_____

_____

Problem Solving

For Exercises 6 and 7, use the information below and the rectangle at the right.

Mrs. Morris plans to tile her front hallway shown at the right.

9 ft
6 ft

6. If each tile is 1 foot long and 1 foot wide, how many tiles will she need?

_____

Mathematical
7. PRACTICE **1** **Keep Trying** Squares of tile come in packages of 6 tiles. How many packages will Mrs. Morris need?

_____

Test Practice

8. Which equation can be used to find the area of the rectangle?

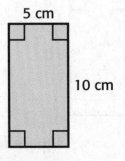
5 cm
10 cm

Ⓐ $5 + 10 = 15$ Ⓒ $5 \times 10 = 50$

Ⓑ $10 - 5 = 5$ Ⓓ $10 \div 5 = 2$

Measurement and Data
3.MD.5, 3.MD.7, 3.MD.7c

CCSS

Hands On
Area and the Distributive Property

Lesson 7

ESSENTIAL QUESTION
How are perimeter and area related and how are they different?

Draw It Tools

The grid shows a rectangle with a length of 6 units and a width of 5 units. If the length of this rectangle increased by 2 units, what would be the new area?

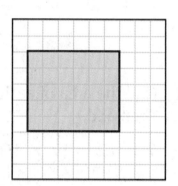

The area of the rectangle is _____ square units. Label this rectangle A.

1 Shade more unit squares so that the length of the rectangle is now increased by 2 units, but the width remains unchanged.

2 Label the additional rectangle formed by what you shaded as Rectangle B. What is the area of Rectangle B? _____

3 Add the areas of rectangles A and B.

$$A = (6 \times 5) + (2 \times 5)$$

$$A = \boxed{} + \boxed{}$$

$$A = \boxed{}$$

The area of the larger rectangle is _____ square units.

Check The length of the larger rectangle is 8 units. The width is 5 units. $8 \times 5 = 40$

The Distributive Property can be used to model the area of a rectangle. Recall that the Distributive Property allows you to decompose one factor.

Try It

Use the Distributive Property to find the area of the rectangle.

 Decompose one factor.

$$12 = 10 + 2$$

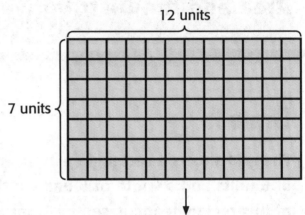

12 units

7 units

10 units   2 units

7 units

 Find the area of each smaller rectangle. Then add.

$$7 \times 12 = (7 \times 10) + (7 \times 2)$$

$$= \boxed{} + \boxed{}$$

$$= \boxed{}$$

So, the area of the rectangle is _____ square units.

Talk About It

1. **Mathematical PRACTICE** ③ **Justify Conclusions** Refer to the second activity. If you decomposed 12 into 9 + 3 instead of 10 + 2, how would that have affected the result?

_____

2. How can the Distributive Property help you find the area of rectangles with greater numbers?

_____

_____

_____

Practice It

Use the Distributive Property to find the area of each rectangle.

3.

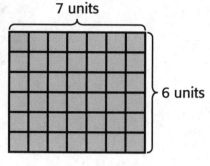

7 units

6 units

4.

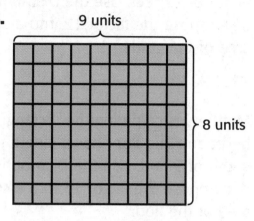

9 units

8 units

$6 \times 7 = (6 \times 5) + (6 \times 2)$

$= \underline{\hspace{1.2cm}} + \underline{\hspace{1.2cm}}$

$= \underline{\hspace{1.2cm}}$

The area is _____ square units.

$8 \times 9 = (8 \times 5) + (8 \times 4)$

$= \underline{\hspace{1.2cm}} + \underline{\hspace{1.2cm}}$

$= \underline{\hspace{1.2cm}}$

The area is _____ square units.

Find the area of each rectangle. Use the Distributive Property to decompose the longer side into a sum. Show your steps.

5.

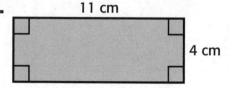

11 cm

4 cm

6.

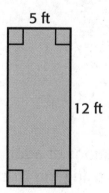

5 ft

12 ft

The area is _____ square centimeters.

The area is _____ square feet.

Apply It

My Work!

I'm feeling fried!

7. Julia is planting vegetables in her rectangular garden. Her garden has a length of 8 feet and a width of 12 feet. Use the Distributive Property to decompose the factor 12 into a sum. Then find the area of the garden.

8. Matthew is carpeting the rectangular floor in his bedroom. The floor has a length of 15 feet and a width of 9 feet. Use the Distributive Property to decompose the factor 15 into a sum. Then find the area of the floor.

9. Mathematical PRACTICE **2** **Reason** Describe three ways to find the area of a rectangle with a length of 9 meters and a width of 4 meters.

10. Mathematical PRACTICE **3** **Find the Error** James needed to find the area of a rectangle with a length of 11 inches and a width of 9 inches. His steps are to the right. Find and correct his error.

$$9 \times 11 = (9 \times 10) + (9 \times 2)$$
$$= 90 + 18$$
$$= 108$$

Write About It

11. How are the operations of addition and multiplication used when finding area using the Distributive Property?

Measurement and Data

3.MD.5, 3.MD.7, 3.MD.7c

CCSS

MY Homework

Lesson 7

Hands On: Area and the Distributive Property

Homework Helper

Need help? ✏ connectED.mcgraw-hill.com

Use the Distributive Property to find the area of the rectangle.

 Decompose one factor.

$11 = 10 + 1$

11 units

7 units

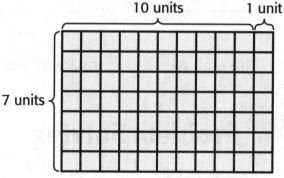

2 Find the area of each smaller rectangle. Then add.

$7 \times 11 = (7 \times 10) + (7 \times 1)$

$= \qquad 70 \quad + \quad 7$

$= \qquad 77$

10 units   1 unit

7 units

So, the area of the rectangle is 77 square units.

Practice

1. Use the Distributive Property to find the area of the rectangle.

$6 \times 9 = (6 \times 5) + (6 \times 4)$

$= \underline{\qquad} \quad + \quad \underline{\qquad}$

$= \underline{\qquad}$

9 units

6 units

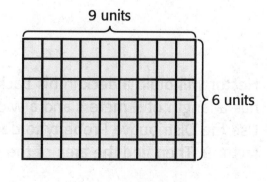

2. Use the Distributive Property to find the area of the rectangle.

$$8 \times 12 = (8 \times 10) + (8 \times 2)$$

= _____ + _____

= _____

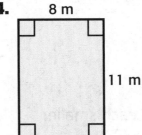

12 units

8 units

Find the area of each rectangle. Use the Distributive Property to decompose the longer side. Show your steps.

3.

9 in.

9 in.

The area is _____ square inches.

4.

8 m

11 m

The area is _____ square meters.

 ## Problem Solving

5. **Mathematical PRACTICE** 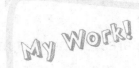 **Identify Structure** Erika is painting a rectangular painting. The painting has a length of 12 inches and a width of 10 inches. Use the Distributive Property to decompose the factor 12. Then find the area of the painting.

_____

My Work!

6. Hector will build a deck in his backyard. The deck has a length of 9 meters and a width of 8 meters. Use the Distributive Property to decompose the factor 9. Then find the area of the deck.

_____

Measurement and Data

3.MD.5, 3.MD.7, 3.MD.7b, 3.MD.7d

CCSS

Area of Composite Figures

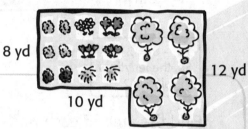

Lesson 8

ESSENTIAL QUESTION
How are perimeter and area related and how are they different?

A **composite figure** is made up of two or more figures. To find the area of a composite figure, decompose the figure into smaller parts.

 Math in My World

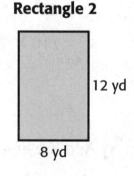

8 yd

12 yd

10 yd

8 yd

Example 1

Shrubs, trees, flowers, and plants can be bought at Mr. Corley's Nursery. What is the area of the nursery's garden at the right?

1 Break the composite figure into smaller parts. Look for rectangles.

2 Find the area of each part.

Rectangle 1   **Rectangle 2**

$A = \ell \times w$ $A = \ell \times w$

    $= 10 \times 8$ $= 12 \times 8$

    $=$ _____ $= (10 \times 8) + (2 \times 8)$ Decompose 12 as 10 + 2.

                            $=$ _____ $+$ _____

                            $=$ _____

Rectangle 1   **Rectangle 2**

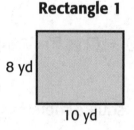

8 yd

10 yd

12 yd

8 yd

The area of Rectangle 1 is _____ square yards. The area of

Rectangle 2 is _____ square yards.

3 Add the areas.

    $80 + 96 =$ _____

The area of the composite figure is _____ square yards.

Example 2

Find the area of the composite figure.

 Break the composite figure into smaller parts. Look for rectangles.

This composite figure can be broken into

_____ rectangles.

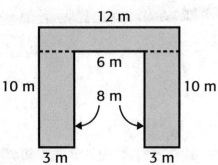

12 m

6 m

10 m   8 m   10 m

3 m   3 m

Rectangle 1

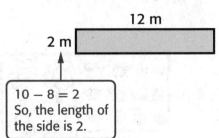

12 m

2 m

$10 - 8 = 2$
So, the length of the side is 2.

Rectangles 2 and 3

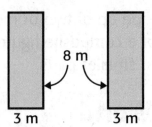

8 m

3 m   3 m

Find the area of each part. Rectangles 2 and 3 are the same size.

Rectangle 1

$A = \ell \times w$

$= 12 \times 2$

$= (10 \times 2) + (2 \times 2)$ Decompose 12.

$= \rule{1.5cm}{0.4pt} + \rule{1.5cm}{0.4pt}$

$= \rule{1.5cm}{0.4pt}$ square meters

Rectangles 2 and 3

$A = \ell \times w$

$= 8 \times 3$

$= \rule{1.5cm}{0.4pt}$ square meters

Add the areas. $24 + 24 + 24 = \rule{1.5cm}{0.4pt}$

The area of the composite figure is _____ square meters.

Guided Practice

1. Find the area of the composite figure. Show your work.

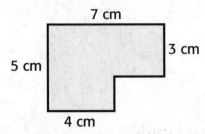

7 cm

3 cm

5 cm

4 cm

The area is _____ square centimeters.

Talk MATH

Refer to Example 1. Find another way to decompose the composite figure.

Name ....

Independent Practice

Find the area of each composite figure. Show your work.

2.

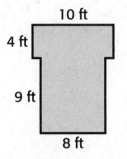

10 ft

4 ft

9 ft

8 ft

The area is _____ square feet.

3.

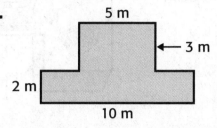

5 m

← 3 m

2 m

10 m

The area is _____ square meters.

4.

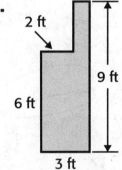

2 ft

9 ft

6 ft

3 ft

The area is _____ .

5.

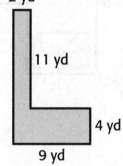

2 yd

11 yd

4 yd

9 yd

The area is _____ .

6. Decompose the composite figure in Exercise 4 a different way. Show the steps you used.

Problem Solving

7. What is the area of the top of the desk?

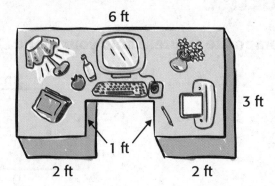

8. Mathematical **PRACTICE** ➊ **Make a Plan** Courtney is playing miniature golf. What is the area of the composite figure?

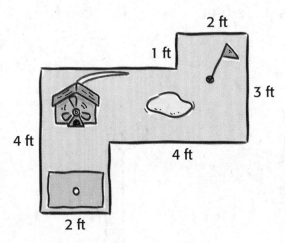

HOT Problems

9. Mathematical **PRACTICE** ➍ **Model Math** Draw and label two composite figures that have the same area but have different perimeters.

10. ❓ **Building on the Essential Question** How is the operation of addition related to finding the area of a composite figure?

Name

MY Homework

Lesson 8

Area of Composite Figures

Homework Helper

eHelp

Need help? connectED.mcgraw-hill.com

Find the area of the composite figure.

1. Break the composite figure into smaller parts. Look for rectangles.

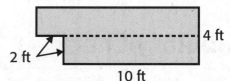

4 ft

2 ft

10 ft

Rectangle 1

12 ft

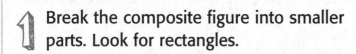

2 ft

10 + 2 = 12
So, this length is 12 feet.

4 − 2 = 2
So, this length is 2 feet.

Rectangle 2

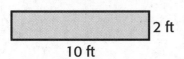

2 ft

10 ft

2. Find the area of each part.

Rectangle 1

$A = \ell \times w$

$= 12 \times 2$

$= (10 \times 2) + (2 \times 2)$ Decompose 12 as 10 + 2.

$= 20 + 4$

$= 24$

Rectangle 2

$A = \ell \times w$

$= 10 \times 2$

$= 20$

The area of Rectangle 1 is 24 square feet.

The area of Rectangle 2 is 20 square feet.

3. Add the areas.

$24 + 20 = 44$

The area of the composite figure is 44 square feet.

Practice

Find the area of each composite figure. Show your work.

1.

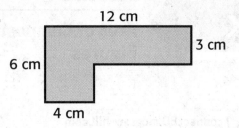
12 cm
3 cm
6 cm
4 cm

2.

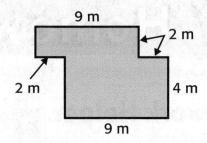
9 m
2 m
2 m
4 m
9 m

The area is _____ square centimeters.   The area is _____ square meters.

Vocabulary Check

3. Draw an example of a composite figure.

Problem Solving

The composite figure shows the floor plan of a bathroom.

4. What is the area of the bathroom floor?

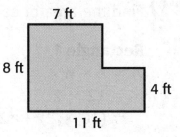
7 ft
8 ft
4 ft
11 ft

5. **Mathematical**
PRACTICE 1 **Plan Your Solution** The floor will be covered in square tiles. If one square tile covers one square foot, how many tiles are needed?

Test Practice

6. What is the area of the composite figure shown?

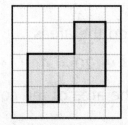

Ⓐ 8 square units   Ⓒ 16 square units

Ⓑ 12 square units   Ⓓ 20 square units

Check My Progress

Vocabulary Check

1. Circle the figure that represents a **composite figure.**
   Explain why the other figures are not composite figures.

_____

_____

_____

2. Circle the **formula** that can be used to find the area of a rectangle.

   $A = \ell + w$ $A = \ell - w$ $A = \ell \times w$

Concept Check

For Exercises 3 and 4, refer to the rectangle shown.

3. Tile the rectangle to find its area. Draw unit
   squares on the rectangle.

   The area is _____ square units.

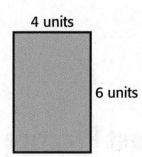

4 units

6 units

4. **Algebra** Write a multiplication equation that can be
   used to find the area of the rectangle without tiling it.

   _____

5. Algebra Find the area of the rectangle. Write a multiplication equation.

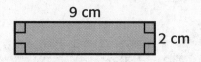

9 cm

2 cm

_____

6. Find the area of the rectangle. Use the Distributive Property to decompose the longer side. Show your steps.

The area is _____ square meters.

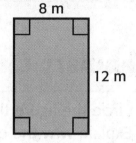

8 m

12 m

Refer to the drawing at the right for Exercises 7 and 8.

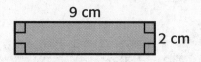 Problem Solving

7. Seth painted the figure at the right on his wall. How many square inches of paint did he use?

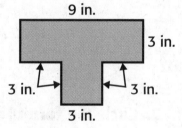

9 in.

3 in.

3 in.

3 in.

3 in.

3 in.

_____

8. Refer to Exercise 8. Decompose the composite figure in a different way to find its area. Show your steps.

Test Practice

9. Which equation can be used to find the area, in square feet, of a rectangle with a length of 8 feet and a width of 4 feet?

 Ⓐ $8 + 4 = 12$ Ⓒ $8 \times 4 = 32$

 Ⓑ $8 - 4 = 4$ Ⓓ $8 \div 4 = 2$

Area and Perimeter

Lesson 9

ESSENTIAL QUESTION
How are perimeter and area related and how are they different?

Two rectangles can have the same area but different perimeters.

 Math in My World 🗨 Tutor

Example 1

Elizabeth will build two fences, one surrounding each garden shown below. How much area does each garden cover? How much fencing will she need for each garden?

 Find the area of each garden.

Garden 1

$A = \ell \times w$

$= 6 \times 2$

$= \underline{\hspace{2cm}}$

6 ft

2 ft

> The gardens have the same area.

Garden 2

$A = \ell \times w$

$= 4 \times 3$

$= \underline{\hspace{2cm}}$

3 ft

4 ft

Find the perimeter of each garden.

Garden 1
The perimeter is 6 + 2 + 6 + 2, or _____ feet.

Garden 2
The perimeter is 3 + 4 + 3 + 4, or _____ feet.

> The gardens have the same area, but different perimeters.

Garden 1 needs _____ feet of fencing.

Garden 2 needs _____ feet of fencing.

Two rectangles can have the same perimeter, but different areas.

Example 2

Draw and label a rectangle that has the same perimeter as the rectangle shown, but a different area.

6 in.

4 in.

 Find the perimeter and area of the rectangle shown.

The perimeter is 6 + 4 + 6 + 4, or _____ inches.

The area is 6 × 4, or _____ square inches. ◄ | Multiply the length by the width.

 Draw and label a rectangle that has a perimeter of 20 inches, but a different area.

 My Drawing

What is the length of the rectangle you drew? _____

What is the width of the rectangle you drew? _____

What is the area of the rectangle you drew? _____

Guided Practice ✓Check

1. Describe the length and width of a rectangle that has the same area as the one below, but a different perimeter.

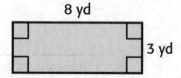

8 yd

3 yd

Copyright © The McGraw-Hill Companies, Inc.   Stockbyte/PictureQuest

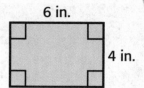 Talk MATH

Refer to Example 2. Describe the length and width of a different rectangle you could have drawn.

Independent Practice

Draw and label a rectangle that has the same area, but a different perimeter, than each rectangle shown.

2. 5 cm

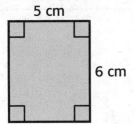

6 cm

3. 4 yd

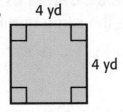

4 yd

Draw and label a rectangle that has the same perimeter, but a different area, than each rectangle shown.

4. 10 ft

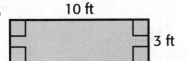

3 ft

5. 4 m

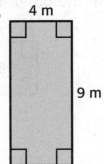

9 m

6. Circle the rectangles that have the same perimeter, but different areas.

8 in

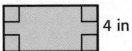

4 in

6 in

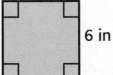

6 in

9 in

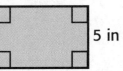

5 in

Problem Solving

William's Windows makes the rectangular windows given in the table. Use this information to solve Exercises 7 and 8.

Window	Length (feet)	Width (feet)
A	6	3
B	5	4
C	9	2

7. **Mathematical PRACTICE 1** **Make Sense of Problems** Which windows will use the same number of square feet of glass?

_____

8. Each window will have a wood border surrounding it. Which windows will use the same amount of wood border?

_____

HOT Problems

9. **Mathematical PRACTICE 3** **Which One Doesn't Belong?** Circle the rectangle that does not belong with the other two. Explain.

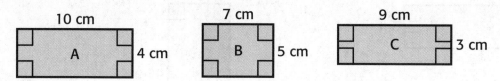

10 cm A 4 cm   7 cm B 5 cm   9 cm C 3 cm

_____

10. **Mathematical PRACTICE 2** **Reason** What is true about the sum of the length and the width for any rectangles with the same perimeter, but different areas?

_____

_____

11. **Building on the Essential Question** How can two rectangles with the same area have different perimeters?

_____

_____

_____

MY Homework

Homework Helper

Need help? connectED.mcgraw-hill.com

Draw and label a rectangle that has the same perimeter as the rectangle shown, but a different area.

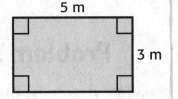

5 m

3 m

1 Find the perimeter and area of the rectangle shown.

The perimeter is 5 + 3 + 5 + 3, or 16 meters.

The area is 5 × 3, or 15 square meters. ◄ Multiply the length by the width.

2 Draw and label a rectangle that has a perimeter of 16 meters, but a different area.

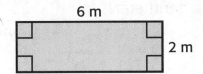

6 m

2 m

The length of the rectangle is 6 meters. The width is 2 meters.

The perimeter is 6 + 2 + 6 + 2, or 16 meters.

The area is 6 × 2, or 12 square meters.

Practice

1. In the space at the right, draw and label a different rectangle that also has a perimeter of 16 meters, but a different area than shown above.

Draw and label a rectangle that has the same area, but a different perimeter, than each rectangle shown.

2.

8 cm

3 cm

3.

9 yd

1 yd

 Problem Solving

David's Dog Pens makes the rectangular dog pens shown in the table. Use this information to solve Exercises 4 and 5.

Dog Pens	Length (feet)	Width (feet)
1	8	6
2	10	4
3	8	5

4. Which dog pens will take up the same area?

_____

5. Which dog pens have the same perimeter?

_____

6. **Mathematical PRACTICE** **1** **Keep Trying** Alexa drew a rectangle with an area of 36 square centimeters. The rectangle she drew has the smallest perimeter possible for this area. What is the length and width of the rectangle she drew?

_____

Test Practice

7. Which rectangle has the same area as Rectangle E, but a different perimeter?

Ⓐ Rectangle A

Ⓑ Rectangle B

Ⓒ Rectangle C

Ⓓ Rectangle D

Rectangle	Length (units)	Width (units)
A	6	6
B	7	6
C	10	3
D	8	5
E	9	4

Measurement and Data

3.MD.5, 3.MD.7, 3.MD.7b, 3.MD.8

CCSS

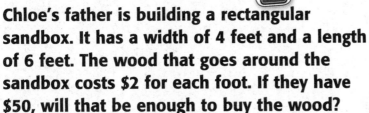

Problem-Solving Investigation

STRATEGY: Draw a Diagram

Lesson 10

ESSENTIAL QUESTION
How are perimeter and area related and how are they different?

Learn the Strategy

Chloe's father is building a rectangular sandbox. It has a width of 4 feet and a length of 6 feet. The wood that goes around the sandbox costs $2 for each foot. If they have $50, will that be enough to buy the wood?

Dig In!

1 Understand

What facts do you know?

• The sandbox is _____ feet by 6 feet.

• The wood costs $ _____ for each foot. They have $ _____.

What do you need to find?

• if $50 is enough money to buy the wood

2 Plan

I can draw a diagram to solve the problem.

3 Solve

Draw a diagram to represent the sandbox.

The perimeter is 6 + 4 + 6 + 4, or _____ feet.

Multiply the perimeter by the cost per foot.

_____ feet × $2 = $ _____

Since $ _____ < $50, they will have enough money.

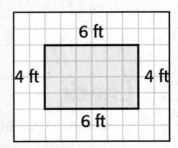

4 Check

Is my answer reasonable? Explain.

Practice the Strategy

A picture frame is 2 inches longer and
2 inches wider than the photo shown.
What is the perimeter of the frame?

4 in.

6 in.

 Understand

What facts do you know?

_____

_____

What do you need to find?

_____

2 Plan

_____

3 Solve

4 Check

Is my answer reasonable?

_____

_____

Apply the Strategy

Solve each problem by drawing a diagram.

1. Phil cut a piece of yellow yarn that was 7 feet long. Kendra cut a piece of red yarn that was 2 feet shorter. William cut a piece of green yarn that was 3 feet longer than the piece Kendra cut. How long was the piece of yarn that William cut?

2. **Mathematical PRACTICE 5 Use Math Tools** Liseta is planting flowers around the outside edge of her rectangular garden. The garden is 12 feet long and 10 feet wide. She will place one flower at each corner and the remaining flowers will be placed 2 feet apart. How many flowers will she plant?

3. A community center is organizing a dance. There are four large columns arranged at the corners of a square. The decorating committee will hang one large streamer from each column to every other column. How many streamers are needed?

4. A cafeteria table has a length of 8 feet and a width of 3 feet. If three tables are pushed together, what is the combined area of the tables?

Review the Strategies

Use any strategy to solve each problem.

• Draw a diagram.
• Solve a simpler problem.
• Use logical reasoning.
• Make an organized list.

5. **Mathematical PRACTICE 5** **Use Math Tools** Madeline has a pink dress, a blue dress, and a yellow dress. She has a black pair of shoes and a white pair of shoes. How many dress and shoe outfits can she make?

My Work!

6. Four friends are in line to see a concert. Greg is last in line. Melanie is before Greg and after Julie. Julie is after Dario. Who is first in line?

7. A restaurant has square dining tables. They will place table settings on each table so that there is a distance of 2 feet from each corner to a table setting. The 3 table settings on each side will be 3 feet apart. What is the perimeter of one table?

8. During one round of a game, Elio, Nida, and Geoffrey each scored 4 points. In round two, they each scored twice as many points. Find the total number of points scored.

9. **Mathematical PRACTICE 1** **Make a Plan** Lanetta will buy balloons for a party. She invited 6 friends from school, 3 friends from soccer practice, and 2 cousins. How many balloons will she need to buy if everyone gets two balloons?

MY Homework

Homework Helper

Need help? connectED.mcgraw-hill.com

Gina's family built a deck in the shape of a hexagon. They placed posts on each outside corner. For a party, they will hang strings of decorative lights from each post to every other post. How many strings of lights are needed?

1 Understand

What facts do you know?
• The deck has six corners.
• One string of lights will be hung from each corner to every other corner.

What do you need to find?
• how many strings of lights are needed

2 Plan

Draw a diagram to solve the problem.

3 Solve

Draw a hexagon.

Draw lines from each corner to every other corner. Each line represents a string of decorative lights.

Count the lines. There are 15 lines drawn.

So, Gina's family needs 15 strings of lights.

4 Check

Is my answer reasonable?

The diagram shows 9 lines inside the hexagon plus 6 lines connecting each side of the hexagon. Since 9 + 6 = 15, the answer is reasonable.

Problem Solving

Solve each problem by drawing a diagram.

Mathematical
1. **PRACTICE** 4 **Model Math** Martina and Charlotte are sharing a pizza. The pizza is cut into eight pieces. Martina ate a quarter of the pizza. Charlotte ate 3 pieces. How many pieces are left?

_____

2. Five friends are having a tennis tournament. Each friend will play the other four friends once. How many matches will be played?

_____

3. A rectangular bedroom floor has an area of 100 square feet and a length of 10 feet. What is the perimeter of the floor?

_____

4. Alexander is riding his bicycle to school. After one mile, he is a third of the way there. How much farther does he have to ride?

_____

5. Marjorie has 28 feet of trim to use as edging on a rectangular blanket she wants to make. What is the length and width of two blanket sizes she could make.

My Work!

Review

Vocabulary Check

Use the word bank to complete each sentence.

area	composite figure	formula
perimeter	square unit	unit square

1. The distance around the outside of a figure is its _____.

2. A unit square has one _____ of area and can be used to measure area.

3. A _____ is an equation that shows the relationship between two or more quantities.

4. _____ is the number of square units needed to cover a figure without overlapping.

5. A square with a side length of one unit is called a _____.

6. A _____ is made up of two or more figures.

Concept Check

Estimate the perimeter of each figure in centimeters. Then measure the perimeter to the nearest centimeter.

7.

Estimate: _____

Actual: _____

8.

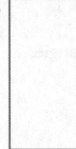

Estimate: _____

Actual: _____

Find the perimeter of each figure.

9.

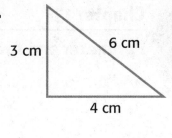

3 cm
6 cm
4 cm

_____ cm

10.

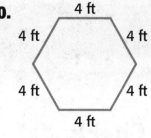

4 ft
4 ft   4 ft
4 ft   4 ft
4 ft

_____ ft

Algebra Find the unknown side length for each figure. The perimeter of each figure is 30 meters.

11.

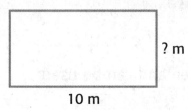

? m
10 m

The unknown is _____ meters.

12.

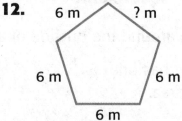

6 m   ? m
6 m   6 m
6 m

The unknown is _____ meters.

Algebra Count unit squares to find the area of each figure. Then write a multiplication equation.

13.

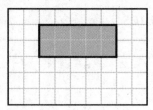

Area: _____

Equation: _____

14.

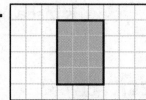

Area: _____

Equation: _____

Find the area of each rectangle.

15.

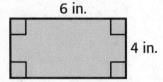

6 in.
4 in.

_____ square inches

16.

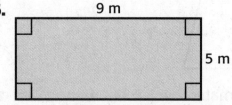

9 m
5 m

_____ square meters

Name _____

Problem Solving

Refer to the figure at the right for Exercises 17–18. The figure represents Kathleen's backyard.

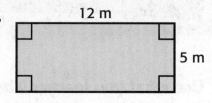

12 m

5 m

17. Use the Distributive Property to find the area of Kathleen's backyard.

_____

18. Tile the rectangle to find its area. What do you notice?

_____

For Exercises 19 through 21, refer to the figure at the right. The figure shows the shape of a room. Each unit square represents 1 square foot.

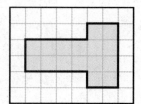

19. What is the area of the room in square feet?

_____

20. What is the perimeter of the room in feet?

_____

21. Decompose the figure into two rectangles to find its area. What do you notice?

_____

Test Practice

22. Find the unknown side length of the figure to the right. The perimeter is 26 inches.

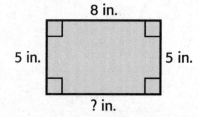

8 in.

5 in.   5 in.

? in.

  Ⓐ 5 inches   Ⓒ 9 inches

  Ⓑ 8 inches   Ⓓ 10 inches

Reflect

Use what you learned about perimeter and area to complete
the graphic organizer.

Units of Measure

Vocabulary

**ESSENTIAL
QUESTION**

How are perimeter and
area related and how
are they different?

Operations Used

Real-World Example

Now reflect on the ESSENTIAL QUESTION Write your answer below.

14 Geometry

Shapes in Our World

Watch a video!

Watch

MY Common Core State Standards

Geometry

3.G.1 Understand that shapes in different categories (e.g., rhombuses, rectangles, and others) may share attributes (e.g., having four sides), and that the shared attributes can define a larger category (e.g., quadrilaterals). Recognize rhombuses, rectangles, and squares as examples of quadrilaterals, and draw examples of quadrilaterals that do not belong to any of these subcategories.

3.G.2 Partition shapes into parts with equal areas. Express the area of each part as a unit fraction of the whole.

Hmm, I don't think this will be too bad!

Standards for Mathematical PRACTICE

1. Make sense of problems and persevere in solving them.
2. Reason abstractly and quantitatively.
3. Construct viable arguments and critique the reasoning of others.
4. Model with mathematics.
5. Use appropriate tools strategically.
6. Attend to precision.
7. Look for and make use of structure.
8. Look for and express regularity in repeated reasoning.

 = focused on in this chapter

Name
....

Am I Ready?

 Check ← Go online to take the Readiness Quiz

Label each shape as a triangle, quadrilateral, pentagon, or a hexagon.

1.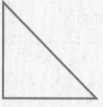

_____

2.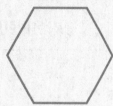

_____

3.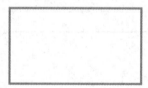

_____

4.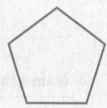

_____

Draw lines to partition each shape.

5. 3 equal parts

6. 2 equal parts

7. Circle the figure that does not belong with the other three. Explain.

_____

Shade the boxes to show the problems you answered correctly.

How Did I Do? → | 1 | 2 | 3 | 4 | 5 | 6 | 7 |

Name ....

MY Math Words

Vocab abc

Review Vocabulary

rectangle	square	triangle

Making Connections
Draw or describe each review vocabulary word.

Draw or describe the shape.

Name the shape.

Shapes

square

rectangle

triangle

What differences and similarities do you notice between each shape?

....

....

....

Copyright © The McGraw-Hill Companies, Inc.   Digital Light Source, Inc.   Photodisc/Getty Images

MY Vocabulary Cards

Lesson 14–1

angle

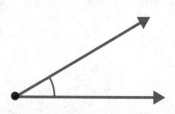

Lesson 14–2

attribute

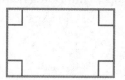

**4 right angles
opposite sides parallel**

Lesson 14–1

endpoint

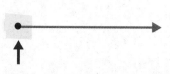

Lesson 14–2

hexagon

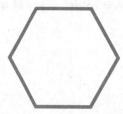

Lesson 14–2

octagon

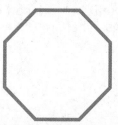

Lesson 14–4

parallel

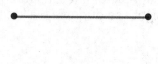

Lesson 14–4

parallelogram

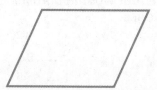

Lesson 14–2

pentagon

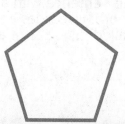

Ideas for Use

- Group common words. Add a word that is unrelated to the group. Then work with a friend to name the unrelated word.

- Design a crossword puzzle. Use the definition for each word as the clues.

A characteristic of a shape.

How can a shape's attributes help you classify it?

_____

_____

_____

Two rays sharing the same endpoint.

Imagine a friend showed you a drawing of two rays. Is the drawing an angle? Explain.

_____

_____

A polygon with six sides and six angles.

Compare a hexagon with a right triangle. Use the space below to draw your comparison.

_____

The point at the beginning of a ray.

Endpoint is a compound word made of the words *end* and *point.* What is another example of a compound word?

_____

Lying the same distance apart from one another.

Name an example of lines that are parallel that you might see during a walk down the street.

_____

A polygon with eight sides and eight angles.

Read and solve this riddle: I am an ocean animal. The prefix *oct-* is a part of my name. I have eight legs. What am I?

_____

A polygon with five sides and five angles.

Complete this sentence with a chapter vocabulary word: Five sides and five angles are called _____ of a pentagon.

_____

A quadrilateral that has both pairs of opposite sides parallel and equal in length.

What is the root word in *parallelogram?* Use it in a sentence.

_____

_____

Lesson 14–2

polygon

Lesson 14–2

quadrilateral

Lesson 14–1

ray

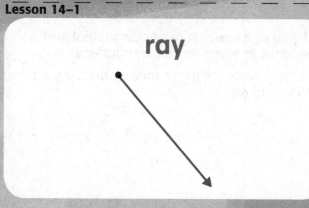

Lesson 14–4

rectangle

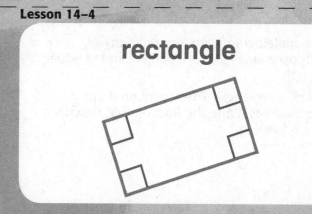

Lesson14–4

rhombus

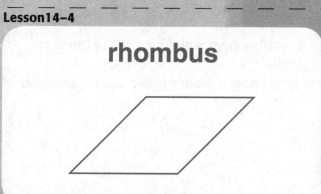

Lesson 14–1

right angle

Lesson 14–3

right triangle

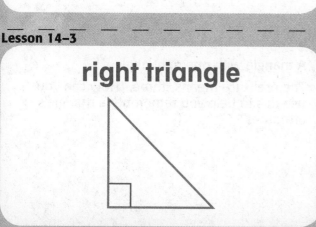

Lesson 14–4

square

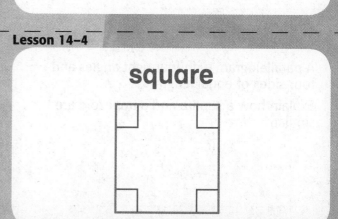

Ideas for Use

- Write a tally mark on each card every time you read or write each word. Challenge yourself to use at least 3 tally marks for each card.

- Draw or write examples for each card. Be sure your examples are different from what is shown on each card.

A shape that has four sides and four angles.

What does the prefix *quad-* mean?

_____

A closed two-dimensional figure formed of three or more straight sides that do not cross each other.

Use a dictionary to find the meaning of the prefix *poly-*. How does it relate to the meaning of polygon?

_____

_____

A parallelogram with four right angles, opposite sides that are parallel and of equal length.

Draw a rectangle. Then partition it into 3 equal areas. Write the fraction that describes your drawing.

A part of a line that has one endpoint and extends in one direction without ending.

Write a sentence using another meaning for the word *ray*.

_____

An angle that forms a square corner.

Give three examples of right angles in a classroom.

_____

_____

A parallelogram with four sides of the same length.

Explain how a rhombus and square are alike.

_____

_____

_____

A parallelogram with four right angles and four sides of equal length.

Explain how a square and a trapezoid are similar.

_____

A triangle with one right angle.

The prefix *tri-* means "three." How can you use this to help you remember a triangle's attributes?

_____

_____

MY Vocabulary Cards

Lesson 14–4

trapezoid

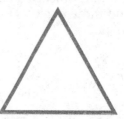

Lesson 14–2

triangle

Lesson 14–1

vertex

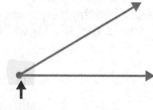

Ideas for Use

- Write the name of a lesson on the front of a blank card. Write a few study tips for that lesson on the back of the card.

- Use the blank cards to draw or write examples that will help you with concepts like the relationship between quadrilaterals and parallelograms.

- -

A polygon with three sides and three angles.

Imagine your friend draws a shape with three sides and three angles. How would you classify the shape?

_____

A quadrilateral with exactly one pair of parallel sides.

Explain how a square and a trapezoid are different.

_____

_____

The shared endpoint where 2 rays meet in an angle.

Vertex can mean "the highest point of something." How is this similar to the math meaning?

_____

MY Foldable

FOLDABLES® Follow the steps on the back to make your Foldable.

✂

Hexagons

Non-examples

Examples

Triangles

Non-examples

Examples

Triangles _____ sides

Quadrilaterals _____ sides

Pentagons _____ sides

Hexagons _____ sides

Octagons _____ sides

Polygons

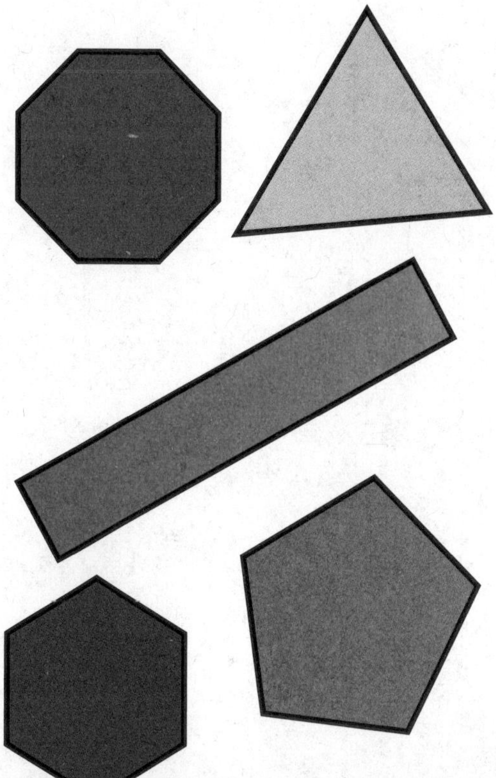

FOLDABLES®
Study Organizer

① Polygons / Triangles
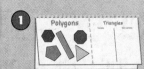
② Polygons / Triangles
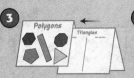
③ Polygons / Triangles
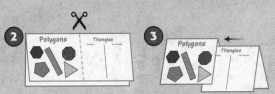
④ Polygons

Quadrilaterals

Non-examples

Examples

Pentagons

Non-examples

Examples

Octagons

Non-examples

Examples

Hands On
Angles

Lesson 1

ESSENTIAL QUESTION
How can geometric shapes help me solve real-world problems?

An **angle** is made when two rays share the same endpoint. A **ray** is part of a line that has one endpoint and extends in one direction without ending. An **endpoint** is the point at the beginning of a ray. The shared endpoint is called the **vertex**.

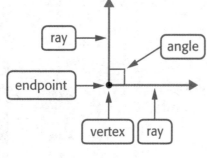

ray

angle

endpoint

vertex   ray

Build It

Use a geoboard and pattern blocks to explore angles.

1 Use a rubber band to make a large square on a geoboard that is similar to the orange pattern block.

2 Use an index card to compare one angle formed by two sides of the square. An angle that forms a *square corner* is called a **right angle**.

Do all four corners of a square form right angles? _____

Try It

1. Use a rubber band to make a large triangle on a geoboard that is similar to the green pattern block.

2. Use an index card to compare one angle formed by two sides of the triangle. This angle is *less than* a right angle.

How many angles in this triangle are less than a right angle? _____

Try It

1. Use a rubber band to create a large hexagon on a geoboard that is similar to the yellow pattern block.

2. Use an index card to compare an angle formed by two sides of the shape. This angle is *greater than* a right angle.

How many angles in the shape are greater than a right angle? _____

Talk About It

1. Can a triangle have two right angles? Explain.

_____

_____

2. **Mathematical PRACTICE 4** **Model Math** Give a real-world example of a right angle.

Name

....

Practice It

Tell whether each angle shown is a *right angle*, *less than* a right angle, or *greater than* a right angle. Use an index card if needed.

3.

4.

....

5.

6.

....

7.

8.

....

9.

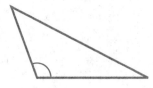

10.

....

Apply It

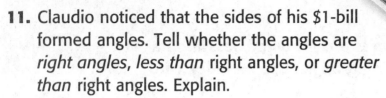

11. Claudio noticed that the sides of his $1-bill formed angles. Tell whether the angles are *right angles*, *less than* right angles, or *greater than* right angles. Explain.

_____

_____

_____

12. Mrs. Wurzer drew four shapes on the board. Circle the shape that appears to have one or more right angles.

13. **Mathematical PRACTICE** 4 **Model Math** Draw three shapes, each showing a different type of angle. Mark each angle and label it.

14. **Mathematical PRACTICE** 2 **Reason** Circle the two angles, in the figure to the right, that appear greater than a right angle. Expain.

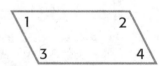

_____

Write About It

15. How can I tell if an angle is a right angle? Explain.

_____

_____

MY Homework

Homework Helper

Need help?  connectED.mcgraw-hill.com

Geoboards and pattern blocks help to explore angles.

1 A rubber band was used to create a large shape on a geoboard that is similar to the red pattern block.

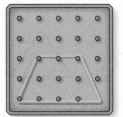

2 An index card was used to compare an angle formed by two sides of the shape. This angle is *less than* a right angle.

There are 2 angles that are *less than* right angles.
There are 2 angles that are *greater than* right angles.

Practice

Tell whether each angle shown is a *right angle, less than* a right angle, or *greater than* a right angle. Use an index card if needed.

1.

2.

3.

_____   _____   _____

Problem Solving

4. **Mathematical PRACTICE 1** **Keep Trying** Draw a time when the hands on the clock make a right angle.

5. **Mathematical PRACTICE 7** **Identify Structure** Mr. West drew four shapes on the board. Circle the shape that appears to have angles that are all less than a right angle.

6. Manny noticed that the sides of his poster on his bedroom wall formed angles. Tell whether the angles are *right* angles, *less than* right angles, or *greater than* right angles. Explain.

_____

_____

_____

Vocabulary Check

Choose the correct word(s) to complete each sentence.

angle   ray   endpoint   vertex   right angle

7. The shared endpoint of two rays is called the _____.

8. An _____ is the point at the beginning of a ray.

9. An angle that forms a *square corner* is called a _____.

Polygons

Lesson 2

ESSENTIAL QUESTION
How can geometric shapes help me solve real-world problems?

A **polygon** is a closed two-dimensional figure formed by three or more straight sides that do not cross each other.

You can classify polygons using one or more of the following attributes. An **attribute** is a characteristic of a figure.

• number of sides   • number of angles

I'm made of many polygons!

 Math in My World   Tools   Watch   Tutor

Example 1

Look at the soccer ball. The blue shape outlined in white is a polygon. Describe and classify the polygon by its attributes.

A **pentagon** is a polygon with 5 sides and 5 angles.

The blue shape outlined in white has _____ sides

and _____ angles.

So, the shape is a pentagon.

Example 2   Tutor

The road sign shown is a polygon. Describe and classify the polygon by its attributes.

An **octagon** is a polygon with 8 sides and 8 angles.

The road sign has _____ sides and _____ angles.

So, the sign is an octagon.

STOP

Key Concept Polygons

Shape	Sides	Angles	Models
triangle	3	3	
quadrilateral	4	4	
pentagon	5	5	
hexagon	6	6	
octagon	8	8	

Talk MATH

What attributes do the shapes in the Key Concept box have in common?

Guided Practice

1. Describe the shape of the sign below. Determine the number of sides and angles. Then classify the shape.

The polygon has _____ sides

and _____ angles.

So, the sign is a(n) _____ .

Name _____

Independent Practice

Describe each shape. Determine the number of sides and angles. Then classify each shape.

2.

_____ sides

_____ angles

This is a(n) _____ .

3.

_____ sides

_____ angles

This is a(n) _____ .

4.

_____ sides

_____ angles

This is a(n) _____ .

5.

_____ sides

_____ angles

This is a(n) _____ .

6.

_____ sides

_____ angles

This is a(n) _____ .

7.

_____ sides

_____ angles

This is a(n) _____ .

Draw an example of each polygon.

8. triangle

9. quadrilateral

CCSS

Problem Solving

10. Bryson pushed a square pattern block and a triangular pattern block together as shown. What new polygon did he create?

_____

11. Classify the polygon that has fewer angles than a quadrilateral.

_____

12. Mathematical **PRACTICE 6** **Explain to a Friend** Explain why each figure below is not a polygon.

_____

_____

HOT Problems

13. Mathematical **PRACTICE 1** **Keep Trying** Draw an example of a figure that is not a polygon. Explain.

_____

_____

14. Mathematical **PRACTICE 7** **Identify Structure** Draw and classify a polygon that has 4 sides with 2 angles that are greater than right angles.

15. **Building on the Essential Question** How do I classify polygons using their attributes?

_____

MY Homework

Homework Helper

Need help? connectED.mcgraw-hill.com

The front of the bird house shown has the shape of a polygon. Describe and classify the polygon.

The polygon has 5 sides and 5 angles.

It is a pentagon.

Practice

Describe each shape. Determine the number of sides and angles. Then classify each shape.

1.

_____ sides

_____ angles

This is a(n) _____ .

2.

_____ sides

_____ angles

This is a(n) _____ .

3. **Mathematical PRACTICE 7 Identify Structure** Classify the polygons that are used to create the figure shown.

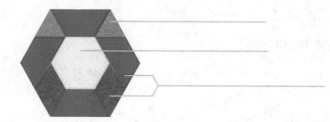

Problem Solving

4. What is another name for a square, other than *polygon*?

_____

5.
Use Math Tools Draw and label the polygon you would get when you fold the hexagon shown, in half along the dotted line.

Polygons help me see!

6. Is the figure shown to the right a polygon? Explain.

_____

_____

Vocabulary Check

Choose the correct word to complete each sentence.

hexagon   polygon   quadrilateral

7. A _____ is a closed two-dimensional figure formed of three or more straight sides that do not cross each other.

8. A _____ is a polygon with 6 sides and 6 angles.

9. A _____ is a polygon with 4 sides and 4 angles.

Test Practice

10. Which of the following figures is a hexagon?

Ⓐ    Ⓑ    Ⓒ 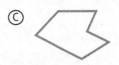   Ⓓ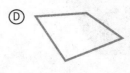

Hands On
Triangles

Lesson 3

ESSENTIAL QUESTION
How can geometric shapes help me solve real-world problems?

Measure It

Measure the sides of each pair of triangles below to the nearest quarter of an inch. Then record the measurements.

Triangle A

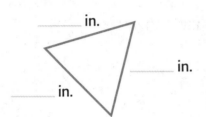

_____ in.

_____ in.

_____ in.

Triangle B

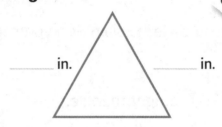

_____ in.

_____ in.

_____ in.

Check out my fin!

Triangle C

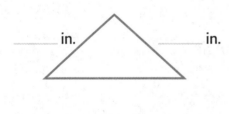

_____ in.

_____ in.

_____ in.

Triangle D

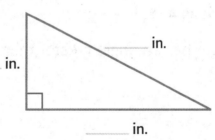

_____ in.

_____ in.

_____ in.

Talk About It

Identify each triangle with what you discovered about the side lengths.

1. No side lengths are the same.   Triangle(s) _____

2. All 3 side lengths are the same.   Triangle(s) _____

3. Exactly 2 side lengths are the same.   Triangle(s) _____

A triangle with one right angle is called a **right triangle**.
Find the right triangle in the photo to the right.

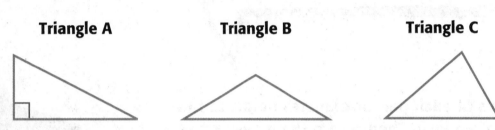

Try It

**Compare the angles of each triangle below. Which triangle
is a right triangle? Circle it.**

Triangle A   **Triangle B**   **Triangle C**

Which triangle did you circle? Explain why you chose that triangle.

_____

So, Triangle _____ is a right triangle.

Talk About It

**Mathematical
PRACTICE 3 Draw a Conclusion** Refer to the activity
above for Exercises 4–5.

4. Which triangle has an angle greater than a right angle?

_____

5. Which triangle has all 3 angles that are less than a
right angle?

_____

_____

6. Explain how a triangle is a special kind of polygon.

_____

_____

Practice It

Measure the sides of each triangle below to the nearest quarter of an inch. Then state the number of sides with equal lengths.

7.

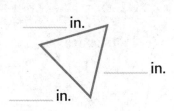

_____ in.

_____ in.

_____ in.

_____ sides

8.

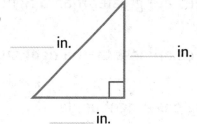

_____ in.

_____ in.

_____ in.

_____ sides

9.

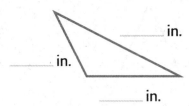

_____ in.

_____ in.

_____ in.

_____ sides

10.

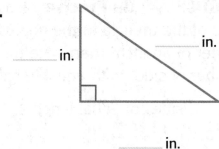

_____ in.

_____ in.

_____ in.

_____ sides

Compare the angles of each triangle. Then circle the correct description.

11.

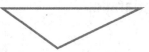

1 right angle

1 angle is greater than a right angle.

12.

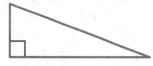

a right triangle

3 angles are less than a right angle.

13.

3 angles are less than a right angle.

1 angle is greater than a right angle.

14.

1 right angle

1 angle is greater than a right angle.

15. Circle the triangles on this page that are right triangles.

Apply It

16. One side of the Egyptian pyramid is in the shape of a triangle. Circle the phrase that best describes the angles of the triangle.

all angles are greater than a right angle

all angles are less than a right angle

one angle is a right angle

17. **Mathematical PRACTICE 6 Be Precise** Measure the sides of the triangle to the nearest quarter of an inch. Then state the number of sides with equal lengths.

_____ sides of equal length

_____ in.   _____ in.

_____ in.

18. **Mathematical PRACTICE 7 Identify Structure** Draw a triangle that is a right triangle. Then draw a triangle that is not a right triangle.

Write About It

19. How are all triangles the same and how can they be different?

_____

_____

_____

Geometry
3.G.1

CCSS

MY Homework

Lesson 3

Hands On:
Triangles

Homework Helper

Need help? connectED.mcgraw-hill.com

Measure the sides of each triangle below to the nearest quarter of an inch. Then state the number of sides with equal lengths.

$\frac{1}{2}$ in.

$\frac{1}{2}$ in.

$\frac{1}{2}$ in.

The triangle has 3 sides with equal lengths.

$\frac{1}{2}$ in. $\frac{1}{2}$ in.

$\frac{3}{4}$ in.

The triangle has 2 sides with equal lengths.

Compare the angles of each triangle. Then describe the triangle using its angles.

The triangle has 1 angle that is greater than a right angle.

The triangle is a right triangle.

Practice

Measure the sides of each triangle below to the nearest quarter of an inch. Then state the number of sides with equal lengths.

1.

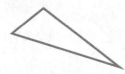

_____ in. _____ in.

_____ in.

_____ sides

2.

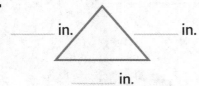

_____ in. _____ in.

_____ in.

_____ sides

Compare the angles of each triangle. Then circle the correct description.

3.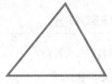

3 angles are less than a right angle

1 right angle.

4.

3 angles greater than a right angle

1 angle is greater than a right angle

Problem Solving

5. **Mathematical PRACTICE** **Be Precise** In billiards, a rack is used to organize billiard balls at the beginning of the game. Measure the sides of the triangle shown. What is the length of each side to the nearest quarter inch?

_____

6. Refer to Exercise 5. How many angles are less than a right angle?

_____

7. How many angles are less than a right angle in the triangle shown at the right?

_____

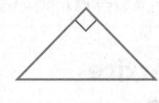

Vocabulary Check

Fill in the missing word.

8. A triangle with one right angle is called a _____ triangle.

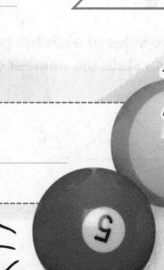

Rack them up!

Quadrilaterals

Lesson 4

ESSENTIAL QUESTION
How can geometric shapes help me solve real-world problems?

Some quadrilaterals have sides that are **parallel**, or equal distance apart.

 parallel

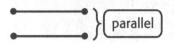 **Math in My World** Tools Watch Tutor

Example 1

The Eiffel Tower shown is in Paris, France. Describe whether any of the sides in the quadrilateral outlined in green are parallel.

The top and _____ sides of the quadrilateral are parallel.

A quadrilateral with *exactly one* pair of parallel sides is a **trapezoid**.

So, the quadrilateral outlined in green is a _____.

Some quadrilaterals have *both* pairs of opposite sides parallel. Those quadrilaterals are called **parallelograms**. Parallelograms have the attributes shown in the table below.

parallelogram

Parallelograms
• both pairs of opposite sides parallel
• opposite sides have the same length
• opposite angles are the same size

There are many kinds of parallelograms. You can classify quadrilaterals, including parallelograms, using the following attributes.

- side lengths
- parallel sides
- right angles

Example 2

Use a centimeter ruler to measure the side lengths of each parallelogram below to the nearest centimeter. Record your results in the table.

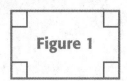

Figure 1

Figure 2

Figure 3

Quadrilateral	Side Lengths (cm)			
	Side 1	Side 2	Side 3	Side 4
Figure 1				
Figure 2				
Figure 3				

Which parallelograms have all sides equal in length? _____

Which parallelograms have four right angles? _____

A **rectangle** is a parallelogram with four right angles.

A **rhombus** is a parallelogram with four equal sides.

A **square** is a parallelogram with four right angles and four equal sides.

Figure 1 is a _____ .

Figure 2 is a rectangle, a rhombus, and a _____ .

Figure 3 is a _____ .

Guided Practice

1. Refer to Example 2. Describe three attributes that Figures 1 and 2 have in common.

_____

_____

Tell why a square is a special kind of parallelogram.

Name ....

Independent Practice

Check all the attributes that describe each quadrilateral.

2.

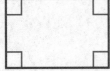

opposite sides are equal length

☐ one pair

☐ both pairs

☐ all four sides same length

opposite sides are parallel

☐ one pair   ☐ both pairs

right angles

☐ 0   ☐ 1   ☐ 2   ☐ 4

This quadrilateral is classified

as a _____ .

3.

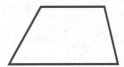

opposite sides are equal length

☐ no pairs

☐ both pairs

☐ all four sides same length

opposite sides are parallel

☐ one pair   ☐ both pairs

right angles

☐ 0   ☐ 1   ☐ 2   ☐ 4

This quadrilateral is classified

as a _____ .

4.

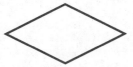

opposite sides are equal length

☐ one pair

☐ both pairs

☐ all four sides same length

opposite sides are parallel

☐ one pair   ☐ both pairs

right angles

☐ 0   ☐ 1   ☐ 2   ☐ 4

This quadrilateral is classified

as a _____ .

5.

opposite sides are equal length

☐ no pairs

☐ both pairs

☐ all four sides same length

opposite sides are parallel

☐ one pair   ☐ both pairs

right angles

☐ 0   ☐ 1   ☐ 2   ☐ 4

This quadrilateral is classified

as a _____ .

Problem Solving

6. The oldest tennis tournament is played at Wimbledon in London. Describe the attributes of the shape of the tennis court. Then classify it.

_____

_____

_____

7. Hilary bought an eraser like the one shown. Classify the quadrilateral formed by the side of the eraser.

8. Kaila is thinking of a quadrilateral. Both pairs of opposite sides are parallel. All four sides are the same length. There are 4 right angles. Draw and label the quadrilateral below.

HOT Problems

9. Circle the quadrilateral(s) that have all the attributes of a rectangle.

parallelogram   rhombus   square   trapezoid

10. Mathematical PRACTICE 7 Identify Structure Circle the quadrilateral(s) that have all the attributes of a parallelogram.

rectangle   rhombus   trapezoid   square

11. ? Building on the Essential Question How can I classify quadrilaterals using their attributes?

_____

_____

MY Homework

Homework Helper

Need help? connectED.mcgraw-hill.com

A tour bus is shown at the right. Describe the attributes of the quadrilateral outlined in yellow. Then classify it.

The quadrilateral has opposite sides that are equal in length and parallel.

It has four right angles.

So, the quadrilateral is a rectangle.

Lady Liberty!

Practice

Describe the attributes of each quadrilateral. Then classify the quadrilateral.

1.

2.

_____

_____

_____

_____

3. Circle the quadrilateral(s) that do *not* have all the attributes of a parallelogram.

   rectangle   rhombus   square   trapezoid

Problem Solving

My home is made of many shapes!

Mathematical PRACTICE 7 **Identify Structure** Check all the quadrilaterals that have the given attributes.

4. Both pairs of opposite sides are parallel.

☐ parallelogram

☐ rhombus

☐ rectangle

☐ square

☐ trapezoid

5. Exactly one pair of opposite sides is parallel.

☐ parallelogram

☐ rhombus

☐ rectangle

☐ square

☐ trapezoid

6. There are four right angles.

☐ parallelogram

☐ rhombus

☐ rectangle

☐ square

☐ trapezoid

7. There are 4 sides that are the same length.

☐ parallelogram

☐ rhombus

☐ rectangle

☐ square

☐ trapezoid

Vocabulary Check

Fill in each blank with a word that makes each sentence true.

8. A square is a parallelogram with _____ right angles and four sides that are the same length.

9. Sides that are the same distance apart are _____ sides.

Test Practice

10. Which of these shapes appears to be a quadrilateral, but *not* a parallelogram?

Ⓐ    Ⓑ    Ⓒ   Ⓓ

Need more practice? Download Extra Practice at ↗ connectED.mcgraw-hill.com

Check My Progress

Vocabulary Check

State whether each sentence is *true* or *false*. If *false*, replace the highlighted word to make a true sentence.

1. A triangle with a right angle is a **right triangle**.

_____

2. A polygon that has 5 sides and 5 angles is a **hexagon**.

_____

3. An **octagon** is made when two rays share the same endpoint.

_____

Concept Check

Describe each shape. Determine the number of sides and angles. Then classify each shape.

4.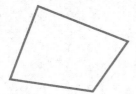

_____ sides

_____ angles

This is a(n) _____.

5.

_____ sides

_____ angles

This is a(n) _____.

6. Describe the attributes of the quadrilateral below. Then classify the quadrilateral.

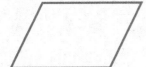

_____

_____

_____

Problem Solving

7. A triangle is a musical instrument. Circle the phrase that best describes the angles of the red triangle.

all angles are greater than a right angle

all angles are less than a right angle

one angle is equal to a right angle

8. Rhonda created two differently-named quadrilaterals with toothpicks. Both quadrilaterals have sides of the same length. What two quadrilaterals did she create?

_____

9. Cole says that all quadrilaterals are polygons, but not all polygons are quadrilaterals. Is he correct? Explain.

_____

_____

10. Three picture frames are on a dresser. Two are shaped like squares and the other is shaped like a trapezoid. How many sides are there in the frames altogether?

_____

Test Practice

11. Mr. Corwin drew four shapes on the whiteboard. Which shape does *not* appear to have a right angle?

Ⓐ

Ⓒ

Ⓑ

Ⓓ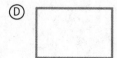

Shared Attributes of Quadrilaterals

Lesson 5

ESSENTIAL QUESTION
How can geometric shapes help me solve real-world problems?

Math in My World

Tools   Tutor

These shapes are all related!

Example 1

The diagram shows how quadrilaterals are related. What attributes are shared by both rectangles and squares?

Quadrilateral
4 sides
4 angles

Trapezoid
One pair of opposite sides are parallel.

Parallelogram
Opposite sides are the same length.
Both pairs of opposite sides are parallel.
Opposite angles are the same size.

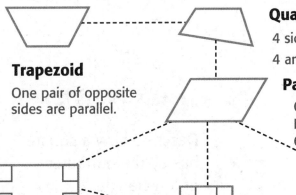

Rectangle
all the attributes of parallelograms plus...
4 right angles

Square
all the attributes of parallelograms plus...
4 sides are the same length
4 right angles

Rhombus
all the attributes of parallelograms plus...
4 sides are the same length

Rectangles and squares share the following attributes.

• Opposite sides are the same _____.

• Both pairs of _____ sides are parallel.

• There are _____ right angles.

Example 2

A parallelogram has opposite sides that are the same length and both pairs of opposite sides parallel. It also has opposite angles that are the same size. Draw an example of a quadrilateral that is *not* a parallelogram.

My Drawing!

Classify the quadrilateral you drew. Explain why it is *not* a parallelogram.

_____

Guided Practice

1. Name one attribute that a square has that a rhombus does not have.

_____

2. Draw an example of a quadrilateral that is *not* a rhombus.

Classify the quadrilateral you drew. Explain why it is *not* a rhombus.

_____

_____

Talk MATH

Describe how a square has all the attributes of a rectangle.

Independent Practice

3. Complete the attributes of a rhombus.

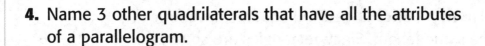

Opposite sides are _____ .

Opposite angles are the _____ size.

The figure has _____ sides that are the same length.

4. Name 3 other quadrilaterals that have all the attributes of a parallelogram.

_____

5. Name another quadrilateral that has all the attributes of a rectangle.

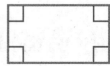

_____

6. Draw an example of a quadrilateral that is *neither* a square, a rectangle, *nor* a rhombus.

Classify the quadrilateral you drew. Explain why it is *not* a square, a rectangle, *or* a rhombus.

_____

_____

Problem Solving

For Exercises 7 and 8, state whether the statement is _true_ or _false_. If false, explain why.

7. All parallelograms have opposite sides that are the same length and parallel. Since rectangles are parallelograms, all rectangles have opposite sides that are the same length and parallel.

_____

_____

8. **Mathematical PRACTICE 3** **Justify Conclusions** All squares have four sides that are the same length. Since rectangles are squares, all rectangles have four sides that are the same length.

_____

_____

HOT Problems

9. **Mathematical PRACTICE 3** **Which One Doesn't Belong?** Circle the quadrilateral that does not belong with the other three. Explain your reasoning.

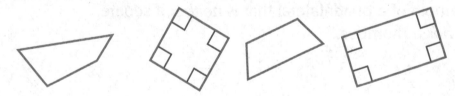

_____

_____

10. **Building on the Essential Question** Explain how a parallelogram is a special kind of polygon.

_____

_____

MY Homework

Homework Helper

Need help? ⟋ connectED.mcgraw-hill.com

The attributes of the quadrilaterals that you learned about
in Lesson 4 were used to create the table.

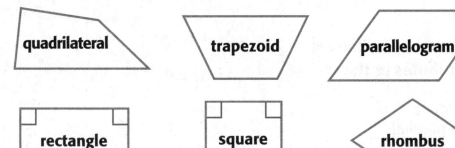

Attribute	Quadrilateral(s)
Both pairs of opposite sides have the same length.	parallelogram, rectangle, square, rhombus
Both pairs of opposite sides are parallel.	parallelogram, rectangle, square, rhombus
Opposite angles are the same size.	parallelogram, rectangle, square, rhombus

Each quadrilateral has 4 sides and 4 angles.

Practice

1. Complete the attributes of a rectangle.

Opposite sides are _____.

Opposite sides are the same _____.

The figure has _____ right angles.

2. Circle the quadrilateral(s) that have all the attributes of
a rectangle.

trapezoid   parallelogram   square   rhombus

Problem Solving

Mathematical
3. PRACTICE 2 Reason State whether the following statement is *true* or *false*. If false, explain why. A trapezoid can also be classified as a parallelogram because it has parallel sides.

_____

_____

_____

For Exercises 4–6, draw a quadrilateral that has the given attributes in the space provided.

My Drawing!

4. opposite sides are parallel

5. four right angles

6. four sides of equal length

Test Practice

7. Which statement about the figures shown below is true?

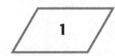

- Ⓐ Figures 1 and 2 are parallelograms.
- Ⓑ Figures 1 and 4 are quadrilaterals.
- Ⓒ Figures 1 and 2 are rectangles.
- Ⓓ Figures 1 and 3 are parallelograms.

Problem-Solving Investigation
STRATEGY: Guess, Check, and Revise

Lesson 6

ESSENTIAL QUESTION
How can geometric shapes help me solve real-world problems?

Learn the Strategy

Julian has five polygons that have a total of 19 sides. Each polygon has 3 or 4 sides. What are the names of the five polygons?

I'm siding with the FACTS!

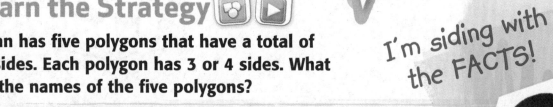

1 Understand

What facts do you know?

• There are _____ polygons and a total of _____ sides.
• The polygons are triangles or quadrilaterals.

What do you need to find?

• Find the _____ of the five polygons.

2 Plan

Guess, check, and revise to solve the problem.

3 Solve

Make a guess, then check. Revise, if needed.

Number of Triangles	Number of Quadrilaterals	Total of Sides	Check
3	2	(3 × 3) + (2 × 4) = 17	too low
2	3	(2 × 3) + (3 × 4) = 18	too low
1	4	(1 × 3) + (4 × 4) = 19	correct

So, Julian has _____ triangle and _____ quadrilaterals.

4 Check

Is my answer reasonable? Explain.

Add the sides of the polygons to check. ____ + ____ + ____ + ____ + ____ = 19

Practice the Strategy

Jolene spent $23 on scrapbook supplies. How many of each supply did she buy?

Scrapbook Supplies	Cost
sticker sheets	$2
packs of paper	$3

 ## 1 Understand

What facts do you know?

_____

_____

_____

What do you need to find?

_____

2 Plan

_____

3 Solve

4 Check

Is my answer reasonable? Explain.

_____

Name

Apply the Strategy

Mathematical
PRACTICE ➊ **Keep Trying** Guess, check, and revise to solve each problem.

large $6

small $5

My Work!

1. A toy store sold $67 worth of stuffed animals. They sold 12 stuffed animals. The prices are shown. How many of each size did they sell?

2. Ed has some hexagon and square pattern blocks. There are a total of 24 sides. He has 5 blocks. How many of each polygon does he have?

3. There are 30 apples in a basket. Half are red. There are 5 more green apples than yellow apples. How many of each color are there?

4. Dawnita has 8 coins. She has quarters, dimes, and nickels. The total is $1. What are the coins?

5. For lunch, Nadia bought two different items. She spent exactly 70¢. What did she buy?

Food	Cost (¢)
box of raisins	35
apple	25
granola bar	45
grilled cheese	85

6. Kira is thinking of two numbers. Their difference is 12 and their sum is 22. What are the numbers?

Review the Strategies

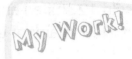

Use any strategy to solve each problem.

- Guess, check, and revise.
- Work backward.
- Draw a diagram.
- Make a table.

7. Tierra ran 4 blocks to her friend's house. Then she ran twice as far to the grocery store. How many blocks did she run in all?

_____

8. Brady is planning a party. He sends invitations to 3 friends from his soccer team, 5 friends from school, and 9 neighbors. Seven friends tell him they cannot come. How many friends will come to the party?

_____

9. Melissa sees 16 wheels on a total of 6 motorcycles and cars. How many motorcycles and cars are there?

_____

10. Mathematical PRACTICE 2 **Use Number Sense** A picnic table has a length of 6 feet and a width of 4 feet. If three tables are pushed together, what is the combined area of the tables?

_____

11. Bishon left home for his friend's house at 6:15 P.M. He arrived at the house 1 hour 35 minutes later. What time did he arrive? Write the time in words and numbers.

_____

12. Mathematical PRACTICE 5 **Use Math Tools** Jerome bought 2 folders and received $1.45 in change in quarters and dimes. If he got 7 coins back, how many of each coin did he get?

_____

My Work!

Squeal!

MY Homework

Homework Helper eHelp

Need help? ↗ connectED.mcgraw-hill.com

Cassandra and Shawnel are the same age. Tonya is 3 years older than Cassandra. If you add all their ages together, the sum is 39. What is the age of each girl?

1 Understand

What facts do you know?

• Cassandra and Shawnel are the same age.

• Tonya is 3 years older than Cassandra.

• The sum of their ages is 39.

What do you need to find?

• Find the ages of each girl.

2 Plan

Guess, check, and revise to solve the problem.

3 Solve

Make a guess, then check. Use what you find to revise.

Cassandra's Age	Shawnel's Age	Tonya's Age	Sum of Ages	Check
10	10	13	33	too low
15	15	18	48	too high
12	12	15	39	correct

So, Cassandra and Shawnel are each 12 years old and Tonya is 15 years old.

4 Check

Is my answer reasonable? Explain.
Add their ages to check. 12 + 12 + 15 = 39

Problem Solving

Mathematical PRACTICE 1 Make a Plan Guess, check, and revise to solve each problem.

1. Mei bought two items. She spent exactly 93¢. What did she buy?

School Supplies	Cost (¢)
eraser	32
pencil	15
pen	20
ruler	61

2. A house has 3 windows that are polygons with a total of 13 sides. Two of the windows are the same shape. The third window has one more side than the first two windows. What specific shapes are the windows?

3. There are 20 crayons in a bag. The crayons are red, yellow, and blue. The number of red crayons is the same as the number of yellow crayons. There are twice as many blue crayons as yellow crayons. How many of each color are there?

4. Dolores bought some new pillows. She bought twice as many green pillows as blue pillows, and 1 less red pillow than green pillows. She bought a total of 9 pillows. How many pillows of each color did she buy?

5. Andrew has a combination of 8 quarters, dimes, and nickels that add up to a value of 95¢. How many of each coin does Andrew have?

Partition Shapes

Math in My World Tools Tutor

Example 1

Victoria will plant 4 different vegetables in her rectangular garden. If she wants to partition the garden into 4 equal sections, what fraction of the garden's area will be used for each vegetable?

1 Draw a rectangle to represent the garden's area.

My Drawing!

2 Partition the rectangle into 4 equal sections. Explain how you partitioned the rectangle.

_____

_____

_____

What unit fraction of the garden's area will be used for each vegetable?

☐ ← section for each vegetable

☐ ← total sections

Example 2

Partition a hexagon into 6 equal sections. What fraction of the hexagon's area does each section represent?

 Draw a hexagon.

My Drawing!

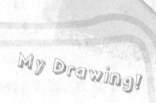

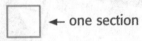 First partition the hexagon into 2 equal sections. Then partition each section into 3 equal sections.

What unit fraction of the hexagon's area does each section represent?

☐ ← one section

☐ ← total sections

Guided Practice [Check ✓]

1. Partition the circle into 4 equal sections. What unit fraction of the circle's area does each section represent?

Each section has an area that is ☐/☐

of the total area of the circle.

Talk MATH
Explain how you would partition a pizza so that you and seven friends each get an equal share.

....

Independent Practice

Partition each figure as indicated. Then write the unit fraction of the figure's area that each equal section represents.

2. 6 equal sections

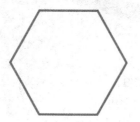

_____

3. 2 equal sections

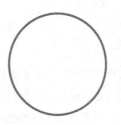

_____

4. 2 equal sections

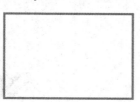

_____

5. 6 equal sections

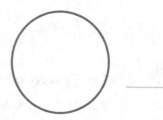

_____

6. 3 equal sections

_____

7. 8 equal sections

_____

8. Draw a square. Partition it into 4 equal sections. What unit fraction of the whole represents the area of each section?

_____

9. Draw a rectangle. Partition it into 5 equal sections. What unit fraction of the whole represents the area of each section?

_____

Problem Solving

10. Loretta and her 3 friends are helping paint a rectangular wall in the school. They decide to partition the wall into 4 equal sections. Each friend will paint one section. What unit fraction of the wall's area will each friend paint?

_____

11. Colton drew a pentagon. He partitioned the pentagon into 10 equal sections. What unit fraction of the whole represents the area of each section?

_____

Don't you love my color?

HOT Problems

12. **Mathematical PRACTICE** 1 **Make Sense of Problems** The area of a rectangular piece of fabric is 24 square inches. The fabric will be partitioned into sections of equal area. Each section will have an area that is $\frac{1}{4}$ of the total area. How many square inches of area will each section have?

_____

13. **Mathematical PRACTICE** 3 **Which One Doesn't Belong?** Circle the figure for which each section represents $\frac{1}{6}$ of its total area.

14. **Building on the Essential Question** How do I partition shapes into parts with equal areas in everyday life?

_____

_____

MY Homework

Homework Helper

Need help? connectED.mcgraw-hill.com

Cassie made a pie to take to the family reunion. If she wants to partition the pie into 8 equal pieces, what fraction of the pie's area will each piece represent?

1 The circle represents the pie's area.

2 Partition the circle into 8 equal sections.

The fraction of the pie's area that each piece represents is $\frac{1}{8}$.

Practice

Partition each figure as indicated. Then write the unit fraction of the figure's area that each equal section represents.

1. 3 equal sections

2. 2 equal sections

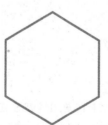

3. 4 equal sections

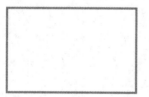

4. 3 equal sections

Problem Solving

5. **Mathematical PRACTICE 5** **Use Math Tools** Nicholas and his 2 friends are raking leaves in his rectangular backyard. They decide to partition the yard into 3 equal sections. Each friend will rake one section. Partition the rectangle into 3 equal sections. Label each section with its unit fraction.

Teamwork!

6. Partition the hexagon into 4 equal sections. What unit fraction of the hexagon's area does each section represent?

_____

7. **Mathematical PRACTICE 2** **Reason** Draw a circle. Partition the circle into six equal sections. What unit fraction of the total area is each section?

_____

Test Practice

8. For art class, each student was given a piece of paper in the shape of a rectangle. Mrs. Brucker asked the students to partition the paper into 8 equal sections. What unit fraction of the paper's area will each section have?

(A) $\frac{1}{2}$

(C) $\frac{1}{6}$

(B) $\frac{1}{3}$

(D) $\frac{1}{8}$

Vocabulary Check

Write each word from the word bank below by its description or example.

angle	attribute	hexagon	octagon
parallel	parallelogram	pentagon	polygon
quadrilateral	rhombus	square	vertex

1.

2. a polygon with 5 sides and 5 angles

3.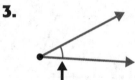

4. a parallelogram with 4 equal sides, but not necessarily 4 right angles

5. a rectangle with 4 equal sides

6.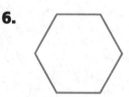

7.

8. a closed figure formed by three or more straight sides that do not cross each other

9.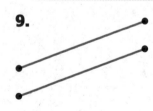

10. Some examples include number of sides, number of angles, and parallel sides.

11. a polygon with 4 sides and 4 angles

12.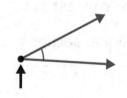

Concept Check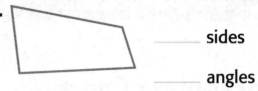

Describe each shape. Determine the number of sides and angles. Then classify each shape.

13.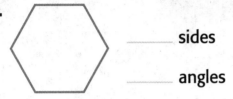

_____ sides

_____ angles

This is a(n) _____.

14.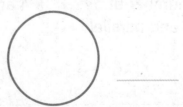

_____ sides

_____ angles

This is a(n) _____.

Describe the attributes of each quadrilateral. Then classify the quadrilateral.

15.

_____

_____

_____

16.

_____

_____

_____

Partition each figure into equal sections as indicated. Then write the unit fraction of the figure's area that each equal section represents.

17. 3 equal sections

_____

18. 6 equal sections

_____

Name _____

Problem Solving

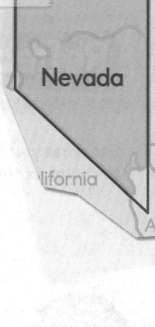

Nevada

lifornia

Ariz

19. The state of Nevada is almost in the shape of a quadrilateral. Complete the attributes of the outline of the state of Nevada.

There is _____ set of parallel opposite sides.

Opposite sides are not equal in length.

Opposite angles are not the same size,

but there appear to be _____ right angles.

20. Karen is thinking of two numbers. Their difference is 9 and their sum is 17. What are the numbers?

_____

21. Naomi cut out two figures from construction paper. One figure is rectangular. The shape of the second figure has all the attributes of the rectangular figure. In addition, it has four sides that are equal in length. Classify the shape of the second figure.

_____

22. Four students were asked to draw a parallelogram. Each drew a differently-named figure, but each was correct. Explain how that can be.

_____

_____

_____

Test Practice

23. Identify the figure that is *not* a trapezoid.

Ⓐ

Ⓒ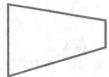

Ⓑ

Ⓓ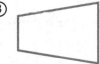

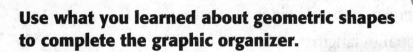

Use what you learned about geometric shapes to complete the graphic organizer.

ESSENTIAL QUESTION

How can geometric shapes help me solve real-world problems?

Real-World Example

Vocabulary

Attributes

Your skills are shaping up!

Now reflect on the ESSENTIAL QUESTION Write your answer below.

_____

_____

Glossary/Glosario

 Vocab ← Go online for the *eGlossary*.

Go to the *eGlossary* to find out more about these words in the following 13 languages:

Arabic • Bengali • Brazilian Portuguese • Cantonese • English • Haitian Creole
Hmong • Korean • Russian • Spanish • Tagalog • Urdu • Vietnamese

Aa

English	Spanish/Español
analog clock A clock that has an *hour* hand and a *minute* hand.	**reloj analógico** Reloj que tiene una manecilla *horaria* y un *minutero*.

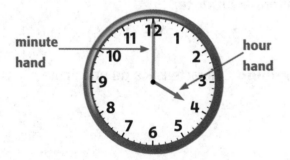

analyze   To break information into parts and study it.

analizar   Separar la información en partes y estudiarla.

angle   A figure that is formed by two *rays* with the same *endpoint*.

ángulo   Figura formada por dos *semirrectas* con el mismo *extremo*.

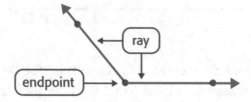

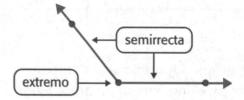

area   The number of *square units* needed to cover the inside of a region or *plane figure*.

área   Cantidad de *unidades cuadradas* necesarias para cubrir el interior de una región o *figura plana*.

area = 6 square units

área = 6 unidades cuadradas

Aa

array   Objects or symbols displayed in rows of the same *length* and columns of the same *length*.

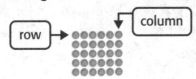

Associative Property of Addition The property that states that the grouping of the addends does not change the sum.

$$(4 + 5) + 2 = 4 + (5 + 2)$$

Associative Property of Multiplication The property that states that the grouping of the *factors* does not change the *product*.

$$3 \times (6 \times 2) = (3 \times 6) \times 2$$

attribute   A characteristic of a shape.

arreglo   Objetos o símbolos organizados en filas y columnas de la misma *longitud*.

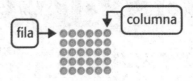

propiedad asociativa de la suma Propiedad que establece que la forma de agrupar los sumandos no altera la suma.

$$(4 + 5) + 2 = 4 + (5 + 2)$$

propiedad asociativa de la multiplicación   Propiedad que establece que la forma de agrupar los *factores* no altera el *producto*.

$$3 \times (6 \times 2) = (3 \times 6) \times 2$$

atributo   Característica de una figura.

Bb

bar diagram   A problem-solving strategy in which bar models are used to visually organize the facts in a problem.

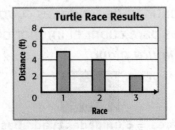

bar graph   A *graph* that compares *data* by using bars of different *lengths* or heights to show the values.

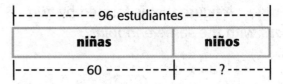

diagrama de barras   Estrategia para la resolución de problemas en la cual se usan barras para modelos de organizar visualmente los datos de un problema.

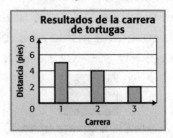

gráfica de barras   *Gráfica* en la que se comparan *los datos* con barras de distintas *longitudes* o alturas para ilustrar los valores.

capacity The amount a container can hold, measured in *units* of dry or liquid measure.

combination A new set made by combining parts from other sets.

Commutative Property of Addition The property that states that the order in which two numbers are added does not change the *sum.*

$$12 + 15 = 15 + 12$$

Commutative Property of Multiplication The property that states that the order in which two numbers are multiplied does not change the *product.*

$$7 \times 2 = 2 \times 7$$

compose To form by putting together.

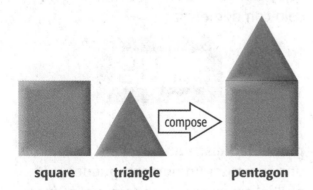

square   triangle   pentagon

composite figure A figure made up of two or more shapes.

capacidad Cantidad que puede contener un recipiente, medida en *unidades* líquidas o secas.

combinación Conjunto nuevo que se forma al combinar partes de otros conjuntos.

propiedad conmutativa de la suma Propiedad que establece que el orden en el cual se suman dos o más números no altera la *suma.*

$$12 + 15 = 15 + 12$$

propiedad conmutativa de la multiplicación Propiedad que establece que el orden en el cual se multiplican dos o más números no altera el *producto.*

$$7 \times 2 = 2 \times 7$$

componer Juntar para formar.

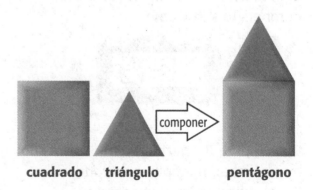

cuadrado   triángulo   pentágono

figura compuesta Figura conformada por dos o más figuras.

data   Numbers or symbols sometimes collected from a *survey* or experiment to show information. *Datum* is singular; *data* is plural.

datos   Números o símbolos que se recopilan mediante una *encuesta* o un experimento para mostrar información.

decagon   A *polygon* with 10 sides and 10 *angles*.

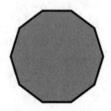

decágono   *Polígono* con 10 lados y 10 *ángulos*.

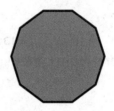

decompose   To break a number into different parts.

descomponer   Separar un número en differentes partes.

denominator   The bottom number in a *fraction*.

In $\frac{5}{6}$, **6 is the denominator.**

denominador   El número de abajo en una *fracción*.

En $\frac{5}{6}$, **6 es el denominador.**

digit   A symbol used to write a number. The ten digits are 0, 1, 2, 3, 4, 5, 6, 7, 8, and 9.

dígito   Símbolo que se usa para escribir un número. Los diez dígitos son 0, 1, 2, 3, 4, 5, 6, 7, 8 y 9.

digital clock   A clock that uses only numbers to show time.

reloj digital   Reloj que marca la hora solo con números.

Distributive Property   To multiply a sum by a number, multiply each *addend* by the number and add the *products*.

$$4 \times (1 + 3) = (4 \times 1) + (4 \times 3)$$

propiedad distributiva   Para multiplicar una suma por un número, puedes multiplicar cada *sumando* por el número y luego sumar los *productos*.

$$4 \times (1 + 3) = (4 \times 1) + (4 \times 3)$$

divide (division) To separate into equal groups, to find the number of groups, or the number in each group.

dividend A number that is being *divided*.

$3\overline{)9}$ 9 is the dividend.

division sentence A *number sentence* that uses the *operation* of *division*.

divisor The number by which the *dividend* is being *divided*.

$3\overline{)9}$ 3 is the divisor.

double Twice the number or amount.

dividir (división) Separar en grupos iguales para hallar el número de grupos que hay, o el número de elementos que hay en cada grupo.

dividendo Número que se *divide*.

$3\overline{)9}$ 9 es el dividendo.

división Enunciado *numérico* que usa la *operación* de *dividir*.

divisor Número entre el cual se *divide* el *dividendo*.

$3\overline{)9}$ 3 es el divisor.

doble Dos veces el número o la cantidad.

Ee

elapsed time The amount of time that has passed from the beginning to the end of an activity.

endpoint The point at the beginning of a *ray*.

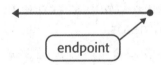

equal groups Groups that have the same number of objects.

equation A *number sentence* that contains an equals sign, =, indicating that the left side of the equals sign has the same value as the right side.

equivalent fractions *Fractions* that have the same value.

$$\frac{2}{4} = \frac{1}{2}$$

tiempo transcurrido Cantidad de tiempo que ha pasado entre el principio y el fin de una actividad.

extremo Punto al principio de una *semirrecta*.

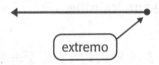

grupos iguales Grupos que tienen el mismo número de objetos.

ecuación *Enunciado* numérico que tiene un signo igual, =, e indica que el lado izquierdo del signo igual tiene el mismo valor que el lado derecho.

fracciones equivalentes *Fracciones* que tienen el mismo valor.

$$\frac{2}{4} = \frac{1}{2}$$

Ee

estimate   A number close to an exact value. An estimate indicates *about* how much.

$$47 + 22 \text{ is about } 70.$$

estimación   Número cercano a un valor exacto. Una estimación indica una cantidad *aproximada*.

$$47 + 22 \text{ es aproximadamente } 70.$$

evaluate   To find the value of an *expression* by replacing *variables* with numbers.

evaluar   Calcular el valor de una *expresión* reemplazando las *variables* por números.

expanded form/expanded notation   The representation of a number as a sum that shows the value of each *digit*.

$$536 \text{ is written as } 500 + 30 + 6.$$

forma desarrollada/notación desarrollada   Representación de un número como la suma que muestra el valor de cada *dígito*.

$$536 \text{ se escribe como } 500 + 30 + 6.$$

experiment   To test an idea.

experimentar   Probar una idea.

expression   A combination of numbers and *operations*.

$$5 + 7$$

expresión   Combinación de números y *operaciones*.

$$5 + 7$$

Ff

fact family   A group of *related facts* using the same numbers.

$5 + 3 = 8$	$5 \times 3 = 15$
$3 + 5 = 8$	$3 \times 5 = 15$
$8 - 3 = 5$	$15 \div 5 = 3$
$8 - 5 = 3$	$15 \div 3 = 5$

familia de operaciones   Grupo de *operaciones relacionadas* que tienen los mismos números.

$5 + 3 = 8$	$5 \times 3 = 15$
$3 + 5 = 8$	$3 \times 5 = 15$
$8 - 3 = 5$	$15 \div 5 = 3$
$8 - 5 = 3$	$15 \div 3 = 5$

factor   A number that is *multiplied* by another number.

factor   Número que se *multiplica* por otro número.

foot (ft)   A customary unit for measuring *length*. Plural is *feet*.

$$1 \text{ foot} = 12 \text{ inches}$$

pie   Unidad usual para medir la *longitud*.

$$1 \text{ pie} = 12 \text{ pulgadas}$$

formula   An *equation* that shows the relationship between two or more quantities.

fórmula   *Ecuación* que muestra la relación entre dos o más cantidades.

fraction A number that represents part of a whole or part of a set.

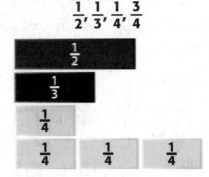

fracción Número que representa una parte de un todo o una parte de un conjunto.

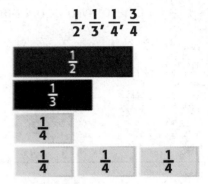

frequency table A *table* for organizing a set of *data* that shows the number of times each result has occurred.

Bought Lunch Last Month	
Name	**Frequency**
Julia	6
Martin	4
Lin	5
Tanya	4

tabla de frecuencias *Tabla* para organizar un conjunto de *datos* que muestra el número de veces que ha ocurrido cada resultado.

Compraron almuerzo el mes pasado	
Nombre	**Frecuencia**
Julia	6
Martín	4
Lin	5
Tanya	4

gram (g) A *metric unit* for measuring lesser *mass.*

gramo (g) *Unidad métrica* para medir la *masa.*

graph An organized drawing that shows sets of *data* and how they are related to each other. Also a type of chart.

bar graph

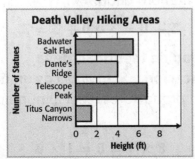

gráfica Dibujo organizado que muestra conjuntos de *datos* y cómo se relacionan. También, es un tipo de diagrama.

gráfica de barras

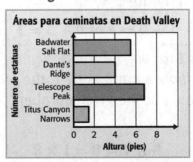

half inch $\left(\frac{1}{2}\right)$ One of two equal parts of an *inch*.

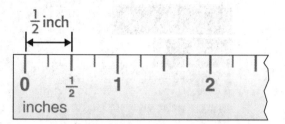

media pulgada $\left(\frac{1}{2}\right)$ Una de dos partes iguales de una *pulgada*.

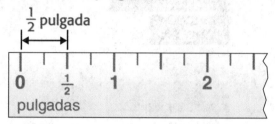

hexagon   A *polygon* with six *sides* and six *angles*.

hexágono   *Polígono* con seis *lados* y seis *ángulos*.

hour (h)   A *unit* of time equal to 60 *minutes*.

1 hour = 60 minutes

hora (h)   *Unidad* de tiempo igual a 60 *minutos*.

1 hora = 60 minutos

hundreds   A position of *place value* that represents the numbers 100–999.

centenas   *Valor posicional* que representa los números del 100 al 999.

Identity Property of Addition   If you add zero to a number, the sum is the same as the given number.

$$3 + 0 = 3 \text{ or } 0 + 3 = 3$$

Identity Property of Multiplication   If you *multiply* a number by 1, the *product* is the same as the given number.

$$8 \times 1 = 8 = 1 \times 8$$

propiedad de identidad de la suma   Si sumas cero a un número, la suma es igual al número dado.

$$3 + 0 = 3 \text{ o } 0 + 3 = 3$$

propiedad de identidad de la multiplicación   Si *multiplicas* un número por 1, el *producto* es igual al número dado.

$$8 \times 1 = 8 = 1 \times 8$$

interpret To take meaning from information.

inverse operations *Operations* that undo each other.

Addition and subtraction are inverse, or opposite, operations.

Multiplication and *division* are also inverse operations.

is equal to (=) Having the same value.

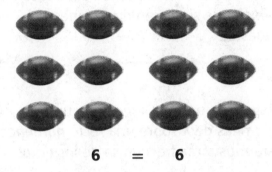

6 = 6

6 is equal to, or the same, as 6.

is greater than > An inequality relationship showing that the value on the left of the symbol is greater than the value on the right.

5 > 3 5 is greater than 3.

is less than < An inequality relationship showing that the value on the left side of the symbol is smaller than the value on the right side.

4 < 7 4 is less than 7.

interpretar Extraer significado de la información.

operaciones inversas *Operaciones* que se anulan entre sí.

La suma y la resta son operaciones inversas u opuestas.

La *multiplicación* y la *división* también son operaciones inversas.

es igual a (=) Que tienen el mismo valor.

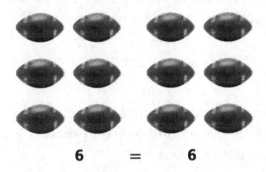

6 = 6

6 es igual o lo mismo que 6.

es mayor que > Relación de desigualdad que muestra que el valor a la izquierda del signo es más grande que el valor a la derecha.

5 > 3 5 es mayor que 3.

es menor que < Relación de desigualdad que muestra que el valor a la izquierda del signo es más pequeño que el valor a la derecha.

4 < 7 4 es menor que 7.

 Kk

key Tells what or how many each symbol in a *graph* stands for.

kilogram (kg) A *metric unit* for measuring greater *mass*.

clave Indica qué significa o cuánto representa cada símbolo en una *gráfica*.

kilogramo (kg) *Unidad métrica* para medir la *masa*.

Kk

known fact A fact that you already know.

hecho conocido Hecho que ya sabes.

length Measurement of the distance between two *points*.

longitud Medida de la distancia entre dos *puntos*.

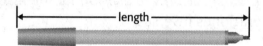
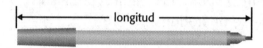

line A straight set of *points* that extend in opposite directions without ending.

recta Conjunto de *puntos* alineados que se extiende sin fin en direcciones opuestas.

line plot A graph that uses columns of Xs above a *number line* to show frequency of *data*.

diagrama lineal Gráfica que usa columnas de X sobre una *recta numérica* para mostrar la frecuencia de los *datos*.

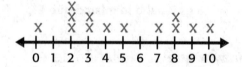
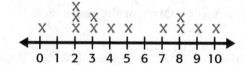

liquid volume The amount of liquid a container can hold. Also known as *capacity*.

volumen líquido Cantidad de líquido que puede contener un recipiente. También se conoce como *capacidad*.

liter (L) A *metric unit* for measuring greater *volume* or *capacity*.

litro (L) *Unidad métrica* para medir el *volumen* o la *capacidad*.

1 liter = 1,000 milliliters

1 litro = 1,000 mililitros

mass The amount of matter in an object. Two examples of *units* of mass are *gram* and *kilogram*.

masa Cantidad de materia en un cuerpo. Dos ejemplos de *unidades* de masa son el *gramo* y el *kilogramo*.

mental math Ordering or grouping numbers so that they are easier to compute in your head.

cálculo mental Ordenar o agrupar números de modo que sean más fáciles de operar mentalmente.

metric system (SI) The measurement system based on powers of 10 that includes *units* such as *meter, gram,* and *liter.*

metric unit A *unit* of measure in the *metric system.*

milliliter (mL) A *metric unit* used for measuring lesser *capacity.*

1,000 milliliters = 1 liter

minute (min) A *unit* used to measure short periods of time.

1 minute = 60 seconds

multiple A multiple of a number is the *product* of that number and any *whole number.*

15 is a multiple of 5 because 3 × 5 = 15.

multiplication An *operation* on two numbers to find their *product.* It can be thought of as repeated addition.

$3 \times 4 = 12$
$4 + 4 + 4 = 12$

multiplication sentence A *number sentence* that uses the *operation* of *multiplication.*

multiply To find the *product* of 2 or more numbers.

sistema métrico (SI) Sistema decimal de medidas que se basa en potencias de 10 y que incluye *unidades* como el *metro,* el *gramo* y el *litro.*

unidad métrica *Unidad* de medida del *sistema métrico.*

mililitro (mL) *Unidad métrica* para medir las *capacidades* pequeñas.

1,000 mililitros = 1 litro

minuto (min) *Unidad* que se usa para medir el tiempo.

1 minuto = 60 segundos

múltiplo Un múltiplo de un número es el *producto* de ese número y cualquier otro *número natural.*

15 es múltiplo de 5 porque 3 × 5 = 15.

multiplicación *Operación* entre dos números para hallar su *producto.* Puede considerar como una suma repetida.

$3 \times 4 = 12$
$4 + 4 + 4 = 12$

multiplicación *Enunciado numérico* que usa la *operación* de *multiplicar.*

multiplicar Hallar el *producto* de 2 o más números.

number line   A line with numbers marked in order and at regular intervals.

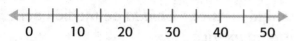

recta numérica   Recta con números marca dos en orden y a intervalos regulares.

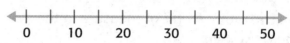

number sentence   An *expression* using numbers and the =, <, or > sign.

$$5 + 4 = 9; 8 > 5$$

enunciado numérico   *Expresión* que usa números y el signo =, <, o >.

$$5 + 4 = 9; 8 > 5$$

numerator   The number above the bar in a *fraction;* the part of the *fraction* that tells how many of the equal parts are being used.

In the fraction $\frac{3}{4}$, 3 is the numerator.

numerador   Número que está encima de la barra de *fracción;* la parte de la *fracción* que indica cuántas partes iguales se están usando.

En la fracción $\frac{3}{4}$, 3 es numerador.

observe   A method of collecting *data* by watching.

observar   Método que utiliza la observación para recopilar *datos.*

octagon   A *polygon* with eight sides and eight *angles.*

octágono   *Polígono* de ocho lados y ocho ángulos.

operation(s)   A mathematical process such as addition (+), subtraction (−), *multiplication* (×), and *division* (÷).

operación   Proceso matemático como la suma (+), la resta (−), la *multiplicación* (×) y la *división* (÷).

parallel (lines)   *Lines* that are the same distance apart. Parallel lines do not meet.

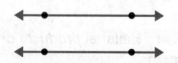

rectas paralelas   *Rectas* separadas por la misma distancia en cualquier punto. Las rectas paralelas no se intersecan.

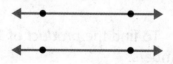

parallelogram   A *quadrilateral* with four sides in which each pair of opposite sides is *parallel* and equal in *length*.

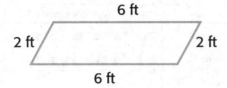

6 ft

2 ft   2 ft

6 ft

paralelogramo   *Cuadrilátero* en el que cada par de lados opuestos son *paralelos* y tienen la misma *longitud*.

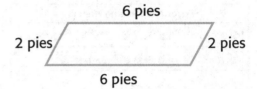

6 pies

2 pies   2 pies

6 pies

parentheses   Symbols that are used to group numbers. They show which *operations* to complete first in a *number sentence*.

paréntesis   Signos que se usan para agrupar números. Muestran cuáles *operaciones* se completan primero en un *enunciado numérico*.

partition   To *divide* or "break up."

separar   *Dividir* o desunir.

pattern   A sequence of numbers, figures, or symbols that follow a rule or design.

2, 4, 6, 8, 10

patrón   Sucesión de números, figuras o símbolos que sigue una regla o un diseño.

2, 4, 6, 8, 10

pentagon   A *polygon* with five sides and five *angles*.

pentágono   *Polígono* de cinco lados y cinco ángulos.

perimeter   The distance around a shape or region.

perímetro   Distancia alrededor de una figura o región.

period   The name given to each group of three *digits* on a *place-value* chart.

período   Nombre dado a cada grupo de tres *dígitos* en una tabla de valor *posicional*..

pictograph   A *graph* that compares *data* by using pictures or symbols.

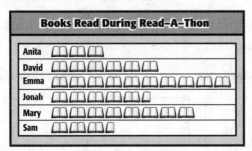

Books Read During Read–A–Thon	
Anita	🔲🔲🔲🔲
David	🔲🔲🔲🔲🔲🔲🔲🔲
Emma	🔲🔲🔲🔲🔲🔲🔲🔲🔲🔲🔲🔲
Jonah	🔲🔲🔲🔲🔲
Mary	🔲🔲🔲🔲🔲🔲🔲🔲
Sam	🔲🔲🔲🔲

pictografía   *Gráfica* en la que se comparan *datos* usando figuras o símbolos.

Libros leídos durante el maratón de lectura	
Anita	🔲🔲🔲🔲
David	🔲🔲🔲🔲🔲🔲🔲🔲
Emma	🔲🔲🔲🔲🔲🔲🔲🔲🔲🔲🔲🔲
Jonah	🔲🔲🔲🔲🔲
Mary	🔲🔲🔲🔲🔲🔲🔲🔲
Sam	🔲🔲🔲🔲

Pp

picture graph A *graph* that has different pictures to show information collected.

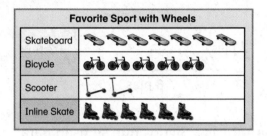

Favorite Sport with Wheels	
Skateboard	
Bicycle	
Scooter	
Inline Skate	

gráfica con imágenes *Gráfica* que tiene diferentes imágenes para ilustrar la información recopilada.

Deporte favorito sobre ruedas	
Monopatín	
Bicicleta	
Patineta	
Patines en línea	

place value The value given to a *digit* by its place in a number.

valor posicional Valor dado a un *dígito* según su lugar en el número.

plane figure A *two-dimensional figure* that lies entirely within one plane, such as a *triangle* or *square*.

figura plana *Figura bidimensional* que yace completamente en un plano, como un *triángulo* o un *cuadrado*.

point An exact location in space.

punto Ubicación exacta en el espacio.

polygon A closed *plane figure* formed by *line* segments that meet only at their *endpoints*.

polígono *Figura plana* cerrada formada por segmentos de *recta* que solo se unen en sus *extremos*.

prediction Something you think will happen, such as a specific outcome of an *experiment*.

predicción Algo que crees que sucederá, como un resultado específico de un *experimento*.

product The answer to a *multiplication* problem.

producto Respuesta a un problema de *multiplicación*.

quadrilateral A shape that has 4 sides and 4 *angles*.

square   rectangle   parallelogram

cuadrilátero Figura que tiene 4 lados y 4 *ángulos*.

cuadrado   rectángulo   paralelogramo

quarter hour One-fourth of an *hour*, or 15 *minutes*.

cuarto de hora La cuarta parte de una *hora* o 15 *minutos*.

quarter inch $\left(\frac{1}{4}\right)$ One of four equal parts of an *inch*.

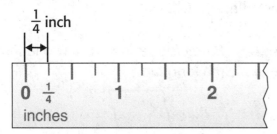

cuarto de pulgada $\left(\frac{1}{4}\right)$ Una de cuatro partes iguales de una *pulgada*.

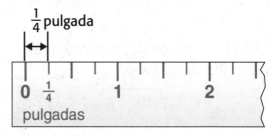

quotient The answer to a *division* problem.

$$15 \div 3 = 5 \quad \longleftarrow \boxed{\text{5 is the quotient.}}$$

cociente Respuesta a un problema *de división*.

$$15 \div 3 = 5 \quad \longleftarrow \boxed{\text{5 es el cociente.}}$$

Rr

ray A part of a *line* that has one *endpoint* and extends in one direction without ending.

semirrecta Parte de una *recta* que tiene un *extremo* y que se extiende sin fin en una dirección.

reasonable Within the bounds of making sense.

razonable Dentro de los límites de lo que tiene sentido.

rectangle A *quadrilateral* with four *right angles*; opposite sides are equal in *length* and are *parallel*.

rectangle *Cuadrilátero* con cuatro *ángulos rectos*; los lados opuestos son de igual *longitud* y *paralelos*.

regroup To use *place value* to exchange equal amounts when renaming a number.

reagrupar Usar el *valor posicional* para intercambiar cantidades iguales cuando se convierte un número.

Rr

related fact(s)   Basic facts using the same numbers. Sometimes called a *fact family*.

$4 + 1 = 5$	$5 \times 6 = 30$
$1 + 4 = 5$	$6 \times 5 = 30$
$5 - 4 = 1$	$30 \div 5 = 6$
$5 - 1 = 4$	$30 \div 6 = 5$

relacionadas   Operaciones básicas que tienen los mismos números. También se llaman *família de operaciones*.

$4 + 1 = 5$	$5 \times 6 = 30$
$1 + 4 = 5$	$6 \times 5 = 30$
$5 - 4 = 1$	$30 \div 5 = 6$
$5 - 1 = 4$	$30 \div 6 = 5$

repeated subtraction   To subtract the same number over and over until you reach 0.

resta repetida   Procedimiento por el que se resta un número una y otra vez hasta llegar a 0.

rhombus   A *parallelogram* with four sides of the same *length*.

rombo   *Paralelogramo* con cuatro lados de la misma *longitud*.

right angle   An *angle* that forms a *square* corner.

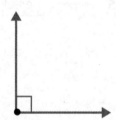

ángulo recto   *Ángulo* que forma una esquina *cuadrada*.

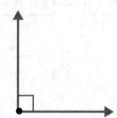

right triangle   A *triangle* with one *right angle*.

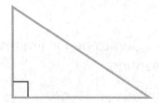

triángulo rectángulo   *Triángulo* con un *ángulo recto*.

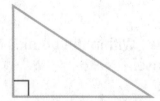

round   To change the value of a number to one that is easier to work with. To find the nearest value of a number based on a given *place value*. 27 rounded to the nearest ten is 30.

redondear   Cambiar el valor de un número a uno con el que es más fácil trabajar. Hallar el valor más cercano a un número con base en un *valor posicional* dado. 27 redondeado a la décima más cercana es 30.

scale A set of numbers that represents the *data* in a *graph.*

square A *plane* shape that has four equal sides. Also a *rectangle.*

square unit A *unit* for measuring *area.*

standard form/standard notation The usual way of writing a number that shows only its *digits,* no words.

537   89   1642

survey A method of collecting *data* by asking a group of people a question.

escala Conjunto de números que representa los *datos* en una *gráfica.*

cuadrado *Figura plana* que tiene cuatro lados iguales. También es un *rectángulo.*

unidad cuadrada *Unidad* para medir el *área.*

forma estándar/notación estándar Manera habitual de escribir un número usando solor sus *dígitos,* sin usar palabras.

537   89   1642

encuesta Método para recopilar *datos* haciendo una pregunta a un grupo de personas.

Tt

table A way to organize and display *data* in rows and columns.

tally chart A way to keep track of *data* using *tally marks* to record the results.

What is Your Favorite Color?	
Color	Tally
Blue	卌 |||
Green	||||

tabla Manera de organizar y representar *datos* en filas y columnas.

tabla de conteo Manera de llevar la cuenta de los *datos* usando *marcas de conteo* para anotar los resultados.

¿Cuál es tu color favorito?	
Color	Conteo
Azul	卌 |||
Verde	||||

tally mark(s)   A mark made to record and display *data* from a *survey*.

thousands   A position of *place value* that represents the numbers 1,000–9,999.

In 1,253, the **1** is in the thousands place.

time interval   The time that passes from the start of an activity to the end of an activity.

time line   A *number line* that shows when and in what order events took place.

Jason's Time Line

Jason born 1999		First day of school 2004	Sister born 2007

1999 2001 2003 2005 2007 2009

trapezoid   A *quadrilateral* with exactly one pair of *parallel* sides.

tree diagram   A branching diagram that shows all the possible *combinations* when combining sets.

triangle   A *polygon* with three sides and three *angles*.

marca de conteo   Marca que se hace para anotar y presentar los *datos* de una.

millares   *Valor posicional* que representa los números del 1,000 al 9,999.

En 1,253, el **1** está en el lugar de los millares.

intervalo de tiempo   Tiempo que transcurre entre de el comienzo y el final de una actividad.

línea cronológica   *Recta numérica* que muestra cuándo y en qué orden ocurrieron los eventos.

Linea Cronológica de Jason

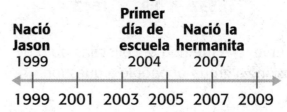

1999 2001 2003 2005 2007 2009

trapecio   *Cuadrilátero* con exactamente un par de lados *paralelos*.

diagrama de árbol   Diagrama con ramas que muestra todas las posibles *combinaciones* al reunir conjuntos.

triángulo   *Polígono* con tres lados y tres *ángulos*.

two-dimensional figure The outline of a shape—such as a *triangle*, *square*, or *rectangle*—that has only *length*, width, and *area*. Also called a *plane figure*.

figura bidimensional Contorno de una figura, como un *triángulo*, un *cuadrado* o un *rectángulo*, que solo tiene *largo*, *ancho* y *área*. También conocida como *figura plana*.

Uu

unit The quantity of 1, usually used in reference to measurement.

unidad Cantidad unitaria, que se usa para referirse a medidas.

unit fraction Any *fraction* with a *numerator* of 1.
$$\frac{1}{2}, \frac{1}{3}, \frac{1}{4}$$

fracción unitaria Cualquier *fracción* cuyo *numerador* es 1.
$$\frac{1}{2}, \frac{1}{3}, \frac{1}{4}$$

unit square A *square* with a side *length* of one *unit*.

cuadrado unitario *Cuadrado* cuya *longitud* de los lados es igual a una *unidad*.

unknown A missing number, or the number to be solved for.

incógnita Número que falta, o el número por el que hay que resolver algo.

Vv

variable A letter or symbol used to represent an *unknown* quantity.

variable Letra o símbolo que se usa para representar una cantidad *desconocida*.

vertex The *point* where two *rays* meet in an *angle*.

vértice *Punto* donde se unen dos semirrectas y forman un *ángulo*.

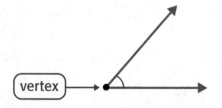

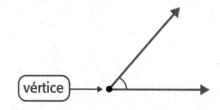

whole number   The numbers
0, 1, 2, 3, 4 . . .

word form/word notation   The form of
a number that uses written words.

6,472

six thousand, four hundred seventy-two

número natural   Los números
0, 1, 2, 3, 4 . . .

forma verbal/notación verbal   Forma
de un número que se escribe en palabras.

6,472

seis mil
cuatrocientos setenta y dos

yard (yd)   A customary unit for
measuring *length*.

1 yard = 3 feet or 36 inches

yarda (yd)   Unidad usual para medir la
longitud.

1 yarda = 3 pies o 36 pulgadas

Zero Property of Multiplication   The
property that states that any number
multiplied by zero is zero.

$$0 \times 5 = 0 \qquad 5 \times 0 = 0$$

propiedad del cero de la multiplicación
Propiedad que establece que cualquier
número multiplicado por cero es igual
a cero.

$$0 \times 5 = 0 \qquad 5 \times 0 = 0$$

Work Mat 1: Thousands Place-Value Chart

thousands	hundreds	tens	ones

Work Mat 2: Number Lines

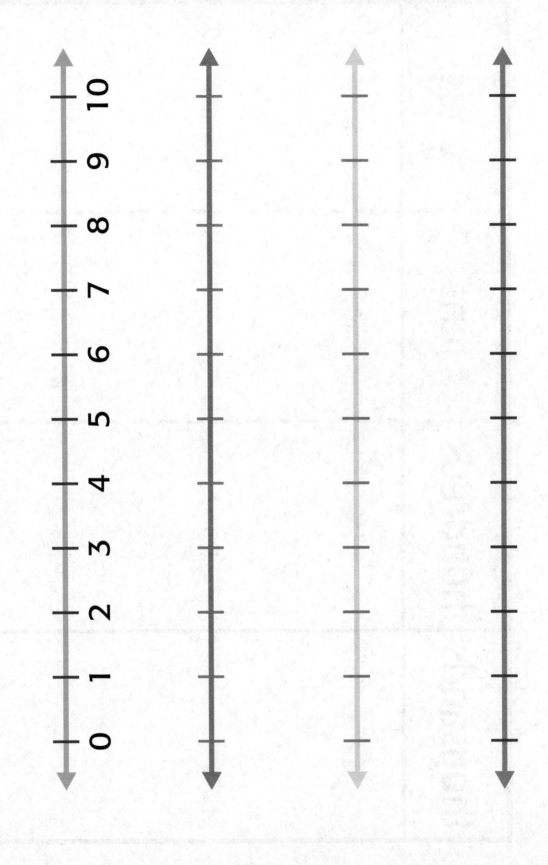

Name

Work Mat 5: Centimeter Grid

Work Mat 6: Bar Diagram

Work Mat 7: Multiplication Fact Table, to 12

✕	0	1	2	3	4	5	6	7	8	9	10	11	12
0	0	0	0	0	0	0	0	0	0	0	0	0	0
1	0	1	2	3	4	5	6	7	8	9	10	11	12
2	0	2	4	6	8	10	12	14	16	18	20	22	24
3	0	3	6	9	12	15	18	21	24	27	30	33	36
4	0	4	8	12	16	20	24	28	32	36	40	44	48
5	0	5	10	15	20	25	30	35	40	45	50	55	60
6	0	6	12	18	24	30	36	42	48	54	60	66	72
7	0	7	14	21	28	35	42	49	56	63	70	77	84
8	0	8	16	24	32	40	48	56	64	72	80	88	96
9	0	9	18	27	36	45	54	63	72	81	90	99	108
10	0	10	20	30	40	50	60	70	80	90	100	110	120
11	0	11	22	33	44	55	66	77	88	99	110	121	132
12	0	12	24	36	48	60	72	84	96	108	120	132	144

Work Mat 8: Algebra Mat

=